# WILLIAM BARRON
## THE VICTORIAN LANDSCAPE GARDENER

T. LIDDLE AND DR P. ROBINSON

AMBERLEY

*This book is dedicated to William Barron, for creating the wonderful gardens and public spaces so familiar to us today, and to all those who have had the fortune of enjoying them.*

First published 2022

Amberley Publishing
The Hill, Stroud,
Gloucestershire, GL5 4EP

www.amberley-books.com

Copyright © T. Liddle and Dr P. Robinson, 2022

The right of T. Liddle and Dr P. Robinson to be identified as the Authors of this work has been asserted in accordance with the Copyright, Designs and Patents Act 1988.

All rights reserved. No part of this book may be reprinted or reproduced or utilised in any form or by any electronic, mechanical or other means, now known or hereafter invented, including photocopying and recording, or in any information storage or retrieval system, without the permission in writing from the Publishers.

ISBN: 978 1 3981 1307 7 (print)
ISBN: 978 1 3981 1308 4 (ebook)

British Library Cataloguing in Publication Data.
A catalogue record for this book is available from the British Library.

Typeset in 10pt on 13pt Celeste.
Typesetting by SJmagic DESIGN SERVICES, India.
Printed in the UK.

*'Although Joseph Paxton is easily the most familiar Victorian landscape gardener, his contemporary William Barron played almost as important a role in forming the image of the landscape gardener as a heroic figure.'*

Brent Elliot, *Victorian Gardens*
(London: B. T. Batsford, 1986)

*'...my chief aim being, to give an impetus to a style of planting in this country, somewhat in accordance with the age of improvement; seeing that facilities are now afforded us which were not presented to our ancestors and predecessors.'*

William Barron, *The British Winter Garden*
(London: Bradbury and Evans, 1852)

*'If you have a garden and a library, you have everything you need'*

Cicero

Walled garden at Elvaston. (T. Liddle)

# Foreword

As I sit here at my desk in a new office, and in a new role as the Head of Tree Collections & Arboriculture, I reflect on the great people who have all played their own special part and made contribution at the Royal Botanic Gardens, Kew. From the great directors of Kew like William Joseph Hooker to the curators like William Jackson Bean, and the plant hunters sending back newly discovered and exciting plants.

One of those great contributors, William Dallimore, became foreman of the arboretum here at Kew and was responsible for the planting of the pinetum. More importantly, he was responsible for the creation and curation of Bedgebury Pinetum, which was Kew's country retreat for the conifer collection to escape the industrial pollution of the time.

William Barron played his own part in the history of Kew Gardens with the creation of the Barron tree-transplanting method, which was revolutionary at the time. But, more important, was his ingenious creation – the Barron Tree Transplanter.

The Royal Botanic Gardens, Kew, purchased one of these new tree transplanters in 1866, nicknamed the 'Devil' by the garden staff due to the injuries that occurred to the staff when using the tree transplanter. The purchase of the tree transplanter allowed for trees that had been grown in the arboretum to be transferred into the landscape within the formal gardens landscape of Kew. This allowed for instant landscapes to be created with semi-mature trees like the three vistas at Kew: Syon Vista, Pagoda Vista and Cedar Vista.

These landscape features of Kew have shaped the arboretum to this day, are enjoyed by millions of visitors, and are of the upmost importance to have been listed as part of Kew's world heritage status.

Kevin Martin<br>
Head of Tree Collections and Arboriculture<br>
Royal Botanic Gardens, Kew

# Preface

*In these days of book and money-making enterprise, time is, as it ever has been, a most valuable commodity; and no one has a right to make a demand on the time of another, without assigning a sufficient reason for so doing.*
William Barron, The British Winter Garden (1854)

In 1854, this is how William Barron started his eponymous book *The British Winter Garden*. Much has changed since he wrote those words, but this point still holds true – time remains a most valuable commodity.

A settlement of some sort has existed at Elvaston for nearly a thousand years – land cultivated and loved by generations of families, farmers, employees, and visitors. For most of its history the space and place flourished, but over the last seventy years time has not been its friend. The Trust is dedicated to preserving Elvaston from the ravages of time, restoring its fabric and breathing new life into a much-treasured estate. It is a long journey to reverse the years of neglect, but one we are committed to. We want Elvaston to last another thousand years.

William Barron is such a key part of the estate's timeline. He worked with the 4th Earl to radically reshape Elvaston's grounds, and with them he reshaped the fundamentals of British gardening, designed revolutionary equipment, perfected grafting, became the first commercial nurserymen, advised government on arboriculture and horticulture and made a significant contribution to the development of public open space. In our view he is an unsung hero, unduly eclipsed by more familiar names such as Capability Brown. However, his contribution is vast, not only paving the way for designers of the quintessential British garden, but also developing and successfully using new, game-changing horticultural techniques.

We decided to write this book on two grounds – our own 'sufficient reasons for doing so'. Firstly, as a testament to the long-lasting legacy and infinite enjoyment of the works of William Barron at Elvaston. Generations of visitors have loved these spaces and we want to ensure they continue to for generations to come. Secondly, to raise up and demonstrate just how important Barron is to the history of British garden design. We believe he is of national significance, not just relevant to our estate, but to all those who have followed him in designing and creating the green spaces we know and love. All the proceeds of the book go to supporting the longevity of Elvaston, its gardens and restoration.

We hope you enjoy this book and in it discover not only the history and importance of Barron's approach and his work at Elvaston, but also something of the man who changed the face of garden design forever. A man who dedicated his life to design and horticulture and to shaping the landscape of our nation.

Peter Robinson and Tamsin Liddle, Chair and Vice Chair<br>Elvaston Castle and Gardens Trust 2021

*The Elvaston Castle and Gardens Trust is an independent charity working with Derbyshire County Council to save the historic Elvaston estate in Derbyshire.*

*You can find out more about Elvaston, the Trust and how to support our work by visiting: www.futureelvaston.co.uk*

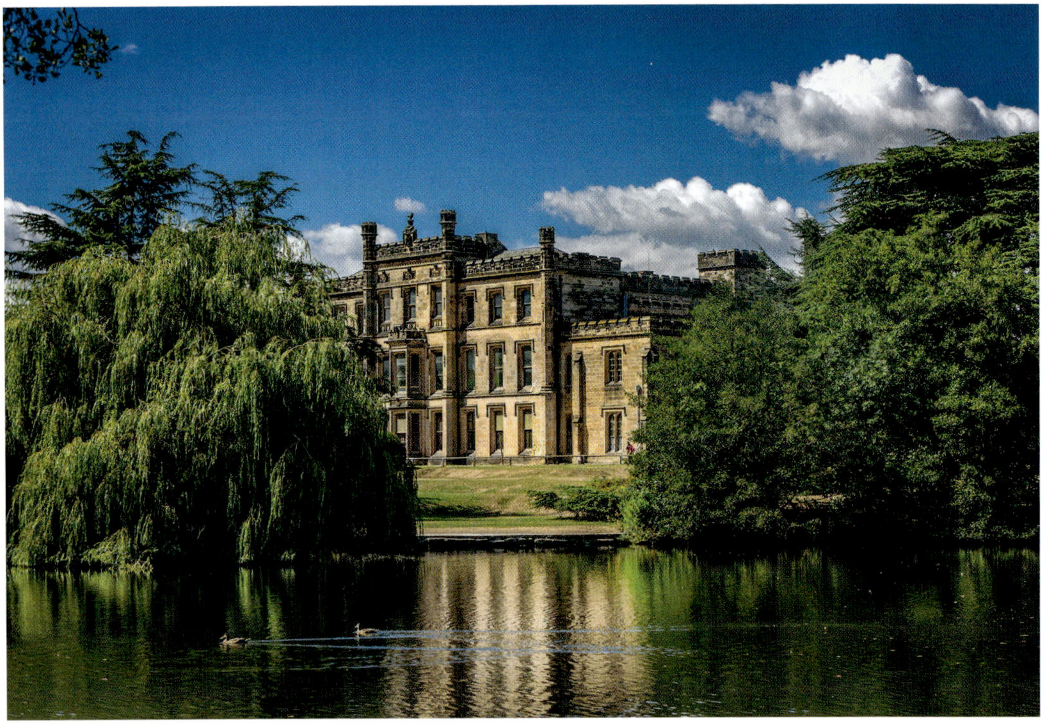

Elvaston Castle lake view. (Derbyshire County Council)

# Contents

| | |
|---|---|
| Introduction | 9 |
| 1 Influences on Victorian Garden Design | 12 |
| 2 The Earls of Harrington | 17 |
| 3 Barron's Early Life and Apprenticeship | 18 |
| 4 The First Years at Elvaston | 24 |
| 5 Elvaston Matures | 28 |
| 6 The British Winter Garden | 31 |
| 7 The Tree-transplanting Machine | 36 |
| 8 Elvaston Opens to The Public | 42 |
| 9 Messrs Barron & Son, Borrowash | 45 |
| 10 The Gardens at Elvaston Today | 50 |
| 11 Barron's Legacy | 52 |
| 12 Private Commissions | 55 |
|     Aqualate Hall, Staffordshire | 56 |
|     Aston Lodge, Aston-on-Trent | 56 |
|     Betley Court, Staffordshire | 57 |
|     Bretby Park, Derbyshire | 60 |
|     Broomfield College, Derbyshire | 60 |
|     Foremark Hall, Derbyshire | 63 |
|     Parkfield Cedars, Derbyshire | 64 |
|     Risley Hall, Derbyshire | 64 |
|     Sennowe Park, Norfolk | 64 |

| | | |
|---|---|---|
| 13 | Public Works | 66 |
| | Abbey Park, Leicester | 67 |
| | Belper Cemetery, Derbyshire | 74 |
| | Brunswick Park, Wednesbury | 76 |
| | Locke Park, Barnsley | 76 |
| | Nottingham Road Cemetery, Derby | 76 |
| | People's Park, Grimsby | 79 |
| | Queen's Park, Chesterfield | 79 |
| | Roath Park, Cardiff | 80 |
| | Victoria Park, Sandwell | 80 |
| | West Park, Macclesfield | 80 |
| | Worcester Pleasure Grounds, Worcestershire | 81 |
| 14 | Other Messrs Barron & Sons Works | 83 |
| 15 | Elvaston Today | 84 |

| | |
|---|---|
| Further Reading and References | 93 |
| Acknowledgements | 94 |
| About the Trust and the Authors | 95 |
|     The Elvaston Castle and Gardens Trust | 95 |
|     Tamsin Liddle | 96 |
|     Dr Peter Robinson | 96 |

# Introduction

Gardens around the world have been a place of enjoyment for thousands of years. In Britain, the beginnings of garden design arrived alongside the Romans and the expansion of their empire. When they landed on these shores in AD 43, they brought with them a completely new style of architecture. The Roman buildings we recognise today from excavations and interpretation boards were a revelation compared to the rustic native structures that preceded them. The Romans believed in aesthetics and refinement, applying those to every area of their lives, including their architectural style. Where buildings in the British Isles were basic and practical, the Romans brought the principles of classical design and beauty to their new settlements, forts, and villas – gardens were a fundamental part of this.

Their gardens were set in the central courtyards of their villa-style homes. This had the dual benefit of protecting plants from the heat and drought, and allowing Roman families to enjoy serene, green spaces at home. Their gardens were adorned with decorative borders, mosaics, statues, and water features, all inspired by earlier Greek and Persian designs. Their plans were simple, elegant, and designed to provide a retreat for relaxing and entertaining – a lasting legacy we benefit from today.

Fast-forward nearly a thousand years, from the end of the Roman empire to the time of the Tudors. Gardens changed from being relaxing spaces to showcases of wealth, designed to amaze and astonish visitors. Private homes and palaces of the rich and royalty were immaculately manicured and carved up into multiple discrete spaces, each with different themes and content. More and more elaborate elements appeared, such as mazes, fountains, and topiary.

Over the next couple of hundred years, however, this trend changed completely. The smaller, elaborate, formal gardens started to give way to sweeping landscapes with large terraces and avenues. As estates became grander the need for an overall plan became more important, and so came the first generation of celebrity garden designers.

In 1716, a young boy was born to a chambermaid and a land agent in rural Northumberland. Lancelot Brown grew up in the countryside at Cambo, watching and learning from his father, John, the estate surveyor at nearby Kirkharle Hall, and his elder brother George, a mason and architect. Unsurprisingly, Lancelot started his working life as an apprentice to the Head Gardener at Kirkharle and remained there until the age of

The 1970s interpretation of a centuries-old topiary tradition at Elvaston. (T. Liddle)

twenty-three. He moved away from home, spreading his wings and earning his first design assignment at Kiddington Hall in Oxfordshire. It was the start of an illustrious career – one that led to him gaining the nickname 'Capability'. By the age of twenty-six he was Head Gardener at the magnificent Stowe House in Buckinghamshire and had already spent time working under the renowned landscape architect William Kent, his predecessor at Stowe. He developed a distinct style of sweeping, undulating grass spaces, punctuated by bundles of mature trees and bodies of water – a style that has become synonymous with the classic image of English countryside. During his lifetime he designed gardens for the best and brightest estates in England: Blenheim Castle, Highclere Castle, Belvoir Castle, Warwick Castle, Althorp – just a few named masterpieces among the hundreds of others he blessed with his signature style.

By the end of the 1700s, after Brown's death, his designs were beginning to be sidelined, with horticultural trends moving towards a new aesthetic favouring messier, wilder spaces and mimicking untouched nature. The pristine, cleaner lines of Brown's designs were increasingly criticised, and his followers started to experiment with this new style. A new hero emerged and Humphry Repton – a lifelong supporter of Brown's work – became his celebrated successor.

Repton was quite different to Brown. He tried his hand at several different trades – journalist, private secretary, artist – failing to find something that suited him. His love of sketching was a constant and in 1788, with a growing family and desperate to provide them with an income, he began to sketch gardens. His childhood friend, who was a well-known botanist, had prompted him to study the natural world; although his knowledge was sparse, he used the little he had learned well. The talented and ambitious Repton rose to fame very quickly, designing for dukes and wealthy landowners alike. Parterres, terraces, and more formal gardens were a familiar feature of his work. He also enjoyed using buildings throughout his landscape design, working with architects such as John Nash and his own son John Adey Repton.

Humphry Repton was a worthy successor to Capability Brown and is noteworthy for bridging the gap between the sweeping expanses of the mid-eighteenth century with the more detailed and ornate Victorian style that developed in the mid-nineteenth century. The late Georgian and early Victorian age was eventful politically, socially and technologically, and it was in this context that Elvaston was remodelled. Britain was confident, buoyed by its success across the globe in establishing a wide and unparalleled empire.

# 1

# Influences on Victorian Garden Design

Born in 1716, Lancelot 'Capability' Brown is acknowledged as Britain's most famous landscape designer, codifying the English landscape style and carrying out work at some 250 country estates. His first appointment as Head Gardener was for Lord Cobham at Stowe. Here he experimented with landscape design, creating sweeping vistas across the Grecian Valley and Elysian Fields while the exceptionally well-connected Lord Cobham openly recommended his Head Gardener to his many aristocratic contacts, including William Pitt, Lord Egremont and George Grenville.

Capability Brown (1716–83).

By 1764 Brown had been appointed as George III's royal gardener, although he also continued to design and manage projects across the country. It is estimated that, between 1761 and 1783 he held contracts valued at some £320,000 (£718 million today).

His landscapes comprised sweeping views, perimeter shelter belts, small tree plantations and lakes. They were simple, uncluttered and designed for a leisure class interested in hunting, shooting and carriage-driving. Describing potential projects as having 'capability', he visited Elvaston Castle in Derbyshire at the behest of the 3rd Earl of Harrington, describing the park as a 'place so flat that there was such a lack of capability in it that he would not meddle with it'. Many of Brown's creations led to the loss of smaller, more formal and often very complex Elizabethan and Jacobean gardens – a natural consequence of changing fashions. There were others who sought to imitate Brown's work, but none achieved the same recognition.

Humphry Repton was perhaps the most successful follower of Brown's approach, focussing on the relationship between the house and the landscape that surrounded it. However, he also reintroduced terracing, flower beds and gravel walks to foreground the landscape and created separate flower gardens.

Repton had recognised in the development of the separate flower garden that fashions were changing. This shift was largely driven by an interest in plant collecting and the resultant need to display these plants. It was also the influence of the Grand Tour and the

Capability Brown landscape at Chatsworth House, Derbyshire. (P. Robinson)

Humphry Repton (1752–1818).

desire to recreate features seen elsewhere in Europe, together with a response to the impact of industrialisation of urban areas.

These factors significantly influenced garden design and saw landscapers return to more traditional forms of planting, including parterres, Italianate terraces and vibrant flower beds, all designed to showcase rare and unusual plants among traditional British varieties.

Away from the formal gardens, arboretums and pleasure grounds were created to display collections of trees while technology (especially boilers and heating systems) allowed gardeners to cultivate rare and tender species in glasshouses and to grow more unusual varieties of fruit. Rockeries were created reminiscent of mountainous European regions and were often planted with alpines. Other areas were left as natural wild gardens.

As garden design evolved, the influences of the Grand Tour also inspired more adventurous landscape gardening. At Biddulph Grange (now a National Trust garden in Staffordshire), James Bateman and his landscape architect Edward Cooke created gardens that were designed to showcase his collection of plants and to celebrate his interest in botany. Different garden spaces, or rooms, were designed with interconnecting architectural devices to take visitors on a journey from Italy to Egypt, to China and to the Himalayas.

Visitors enter one garden through a cottage and emerge from the pyramids of Egypt. In other areas a rockery passage gives way to a Chinese garden (shown opposite) and provides a route through to a stumpery (log and root formations constructed as a fernery).

The Chinese Garden at Biddulph Grange. (P. Robinson)

The gardens are surrounded by an extensive tree collection and feature rhododendrons, summer bedding displays, a stunning Dahlia Walk and the oldest surviving golden larch in Britain. Bateman's interests extended to geology, theology and botany and he was recognised as a collector and scholar of orchids. A contemporary of William Barron, it is noted that much of the evergreen planting at Biddulph was almost certainly inspired by Elvaston.

While this book is not written to introduce the many significant Victorian gardeners of the time, it is important to note the importance of Joseph Paxton as another of the most influential gardeners of the period. Originally employed as a gardener at Chatsworth House, Paxton's skills saw him rise through the ranks to become superintendent in 1826. However, Paxton was also an ambitious architect and created the famous iron and glass conservatory (the site of the maze today) and a lily house. When, in 1850, Paxton proposed a prefabricated iron and glass structure for the Crystal Palace (to house the Great Exhibition), his design replaced the original winning design. This huge structure was completed in six months, and he was knighted in 1851. A keen architect, Paxton also designed many houses and laid out several public parks.

The role of Victorian gardeners in society should not be underestimated. Several were architects, researchers, academics, and writers. John Claudius Loudon, for example, was a Scottish gardener who went on the Grand Tour twice to learn and develop his skills. He was an avid botanist and a respected author and continued his work despite being crippled by a combination of rheumatism and arthritis and having lost his right arm after an operation to fix a break went wrong.

Many of the great Victorian gardeners have come to be described as 'hero' gardeners, their work earning them the same status that today's TV gardeners enjoy. Many of the great Victorian gardens have been lost, sometimes as a result of early twentieth-century landscaping, but many gardens also disappeared with the many country houses that were lost during the great economic turmoil after the two world wars – two great houses a week were being demolished during the 1950s.

Unlike their Georgian predecessors, however, the recognition of the importance of public open space in towns and cities during the Victorian period means that many gardeners of the time worked on both private and public gardening works. While William Barron may have spent much of his life at Elvaston, he too should be counted among these hero gardeners for his impact extends beyond the boundaries of the Elvaston estate. His legacy lives on at Elvaston as well as in the grounds of some surviving country houses and in many public parks across the country, their layouts often unchanged from Barron's original vision.

# 2

# The Earls of Harrington

Born in 1780, Charles Stanhope, Lord Petersham, entered the world into a wealthy, landowning family and the son of another Charles Stanhope, 3rd Earl of Harrington. The 3rd Earl was a popular, successful army officer and diplomat who travelled to North America during the American War of Independence as well as Jamaica, Vienna, and Berlin. The heir to the Harrington earldom, Lord Petersham was given an excellent education at Eton and entered the army, like many of his aristocratic contemporaries, serving in the Coldstream Guards. At the age of thirty-two, he was appointed as a Gentleman of the Bedchamber to George III – a role where he was a companion to the king, helping him dress and controlling access to his bedchamber and wardrobe.

This role aligned very much to his interests: he was a dandy, wealthy enough to enjoy the high life without a financial care in the world. He collected snuff boxes and different types of tea, and he designed his own clothes, even setting a trend for his own design of both a Harrington coat and a Harrington hat. He enjoyed the high life in London – its theatricality, variety of life and pleasures – supposedly not emerging from his London house until 6 p.m. each day, ready to enjoy the evening's entertainment. His home was Harrington House, a mansion situated amid the large swathe of Kensington that the family owned; their influence is still evident today in London street names such as Elvaston Place, Petersham Mews and Stanhope Gardens. 'Beau' Petersham, as he was known, was a handsome man about town, and it was not until he was fifty-one, a year after he inherited the Earl Harrington title, that he married.

Maria Foote was a very popular actress of the time. With light brown hair and a pretty face, she charmed the theatregoers of the day with her acting skills and musical talents. By the time she met Earl Harrington she already had two children from a relationship in 1816 with Colonel William Berkeley and had taken a previous lover to court over a broken promise of marriage. Fearless and independent, she won her case and made her own way in the world. Her marriage to Beau on 7 April 1831 was beset with challenges. In the 1800s, an aristocrat marrying an actress was looked down upon by society and the newlyweds were shunned by their social circle. While they spent some time at their London home, Elvaston was their main retreat and the 4th Earl set about transforming the estate to create a magical and interesting wonderland away from society's prying eyes.

# 3
# Barron's Early Life and Apprenticeship

On 7 September 1805, in the rural Scottish county of Berwickshire, John Barron and his wife Betsy Johnson welcomed a son into the world – William. He grew up in Eccles, a small village in the Scottish Borders – an agricultural area, where his father was a gardener. He was afforded a good education and had an aptitude for mathematics and languages, learning Latin, French, Greek, and Hebrew. Despite the many opportunities his intellect would have attracted, it seems his father's profession was what truly inspired young William, who developed a keen interest in horticulture and chose it as his first steps into the world of work.

William Barron, by an unknown artist.
Oil on panel, nineteenth century.
(Derby Museums Collection)

He served his apprenticeship on the Blackadder estate in Berwickshire. He spent three years on the estate, from the age of fifteen, which had been in possession of the Home family since 1671. Its history is intertwined with the Borders' feuds and its lands had seen bloodshed and warfare for centuries. By the 1800s, there was peace and at around the time Barron was there, peach trees and an English Yew were planted among several types of laurels, smaller trees, holly, and rhododendrons. Blackadder provided the formative practical experience of how to manage a formal garden for Barron, which proved invaluable for his next challenge. Blackadder House was demolished in 1925 after the government refused to pay for damages done by the army following the requisition of the property during the First World War.

At the age of eighteen, William Barron was awarded his first managerial role – and it was a significant one. The glasshouses at the Royal Botanical Gardens in Edinburgh were home to many varied plant species and were the jewel in the Scottish capital's horticultural crown. It was a place for academics to marvel at the wild and exotic specimens and home-grown favourites alike, and to study the medicinal possibilities of the many varieties.

He worked there under the watchful eye of two very influential horticulturalists – Robert Graham (1786–1845) and William McNab (1780–1848). Graham was the 3rd Regius Keeper of the Botanical Gardens and Professor of Medicine and Botany at the city's university, and William McNab was the Botanical Garden's Head Gardener and Curator. Between them, they will have imparted a vast wealth of knowledge on young Barron as he refined his management and horticultural skills. While in Edinburgh, he also had the fortune to be able to attend lectures at the university, studying, among other subjects, botany, mechanics,

Blackadder House, Berwickshire.

McNab map – The Royal Botanic Gardens Edinburgh, as Barron would have known them.

and chemistry. There is no doubt that his time in the city would have set the young man alight in terms of experiences and knowledge, with the expertise and guidance that surrounded him providing a formidable basis on which he built his career as one of the foremost Victorian gardeners.

In 1827, people were starting to take notice of Barron. At the age of twenty-two he was called upon to by the Duke of Northumberland to plant the new conservatory at Syon House in Middlesex. The Great Conservatory, now Grade I listed, was ground-breaking: the first structure of its type to be built using metal and glass, and on a scale previously not achieved. It still stands today, a testament to the longevity and beauty of the horticultural revolution.

The next step for Barron shaped his career and firmly secured him a place in the history books. In 1830, he was invited to Elvaston. He had been recommended to the 4th Earl of Harrington, Charles Stanhope, by his previous employer at the Botanic Gardens, William McNab. A key to Stanhope's acceptance of this recommendation was Barron's background. Three key elements cemented his position. Firstly, he benefitted from the availability and comparative quality of the Scottish education system – parish schools were prevalent in the lowlands and were generally available to all. Secondly, the thoroughness and focus with which Scottish garden apprentices learned their trade was unparalleled in England. They tended to live on-site with their colleagues, away from their families and other distractions.

Derby Midland Railway Hotel
Monday May

Dear Mr MacNab
I came here on Saturday & went to Elvaston Gardens where I received every attention possible from Mr Barron the agreeable & zealous former of these wonderful gardens. He made me aware of some important observations he has made which are not generally known 1. That young trees & plants are permanently injured

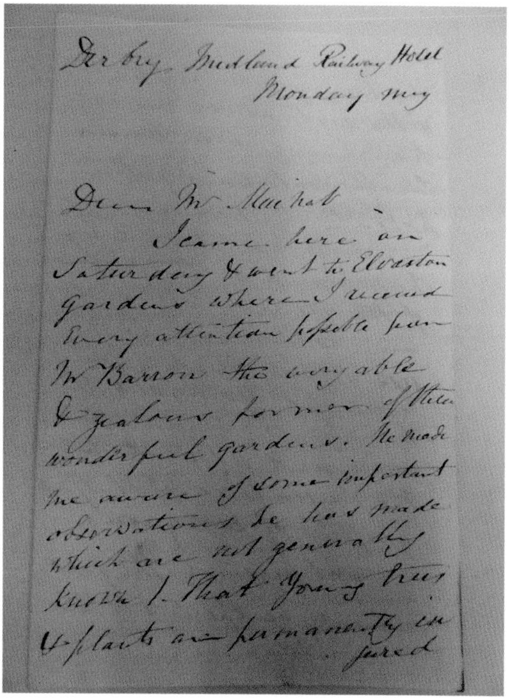

*Right and below*: A letter from McNab referring to Barron. (© Royal Botanic Gardens Edinburgh)

by being grown in pots & the roots never recovering from having been unnaturally contorted by the pots – 2 That evergreens are best transplanted by being placed on the surface of the ground without any hole or pit. 3 The use of young plants of Green Variegated Holly box & to make ornamental divisions & adjuncts to the larger trees & skilful grafting of evergreens

At present the Garden can only be opened to those who come to purchase plants & I left £2 for plants which Mr Barron was to select and send either to you or Mr Evans as he had an opportunity hearing that you or Mr Evans would be kind enough to take charge of them them for me

Returning you many thanks for seeing the distinguished garden at Elvaston Castle to such advantage I remain very sincerely

John A Murray

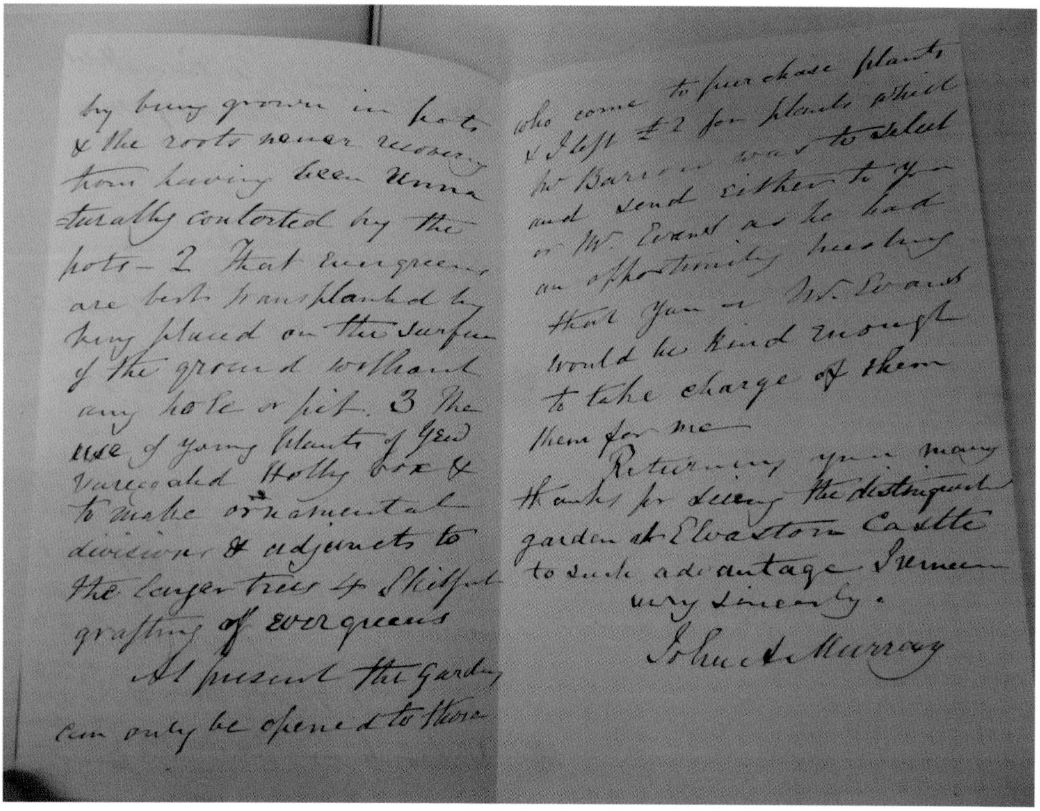

The Great Conservatory, Syon Park, Brentford. (Peter Trimming)

Just as some student's leave home and travel to university today, the apprentices also flew the nest and learned to stand on their own two feet. The nature of the estates they worked on meant they were far from any diversions that may have disrupted their concentration. Finally, education happened not only in the day during their gardening work, but also in the evenings, when the more senior staff would tutor the young in garden and design drawing and plant names. This was often supplemented by access to the vast libraries the great estates were fortunate to possess. The environment in which they had to work was also difficult and required skill to manage – the tougher terrain and inclement weather of Scotland. All this added together to make the Scottish Gardener second to none and William Barron was no exception.

On top of this, Barron had an additional string to his bow: he had carved his early career at the epicentre of the horticulture during the Scottish Enlightenment. It lasted through the eighteenth and early nineteenth centuries and was a time of investigation and exploration into the sciences, politics, architecture, medicine, and many aspects of the natural world such as biology, botany, zoology, agriculture and horticulture. Intellectual debate abounded and societies sprang up in Edinburgh and in ancient universities (Edinburgh, Glasgow, St Andrews, and Aberdeen) dedicated to advancing human knowledge. The examples of its lasting legacy are many.

During this time, the Scottish economist Adam Smith (1723–90) wrote the first modern economics book, commonly known as *The Wealth of Nations* in 1776. It remains a fundamental part of economics studies today. Literature blossomed with the works of James Boswell (1740–95), Robert Burns (1759–96) and Hugh Blair (1718–1800). Between them they produced accounts of travels across Europe, biographies, editions of Shakespeare, lectures on rhetoric and Christian morality, and a canon of famous Scottish poems.

Medicine took a real leap forward during the Enlightenment, and it was this that led to the advancement of horticulture. The two had been inextricably linked for centuries, with plants forming the backbone of treatment for all manner of ailments. As the intellectual investigation of this period grew, so did the number of physicians – both as students learning at the ancient universities and as graduates plying their skills across the country.

As the home of an ancient university with a strong history in medical research, the city's botanical gardens were considered a masterpiece, and still exist today. Barron learned about, nurtured and managed both its specimens and staff. He had the skill to enable a multitude of varieties to co-exist and survive together and also co-ordinate the logistics that enabled it. There were few people with this breadth and depth of skill, particularly at such a young age. It was this combination that led him to being invited to take up the role of Head Gardener at Elvaston and asked to make the gardens 'second to none' by the 4th Earl.

# 4

# The First Years at Elvaston

*No occupation is so delightful to me as the culture of the earth, and no culture comparable to that of the garden.*

Thomas Jefferson

Barron arrived at Elvaston in 1830, the year during which William IV succeeded his brother George IV as king of the United Kingdom. Further afield, the Book of Mormon was published in New York, France invaded Algiers, the Belgian Revolution began, Hector Berlioz's famous *Symphonie fantastique* was premiered in Paris and the Liverpool & Manchester Railway – the first scheduled passenger railway that operated solely using steam power – opened to the public.

He found at Elvaston a flat landscape, broken by two avenues planted on the recommendation of Capability Brown when he met the 4th Earl's grandfather. Brown had seen no merit in attending to the grounds and having presented the 2nd Earl with six cedars of Lebanon to start the avenues, decided he wanted no more to do with the project. Barron determined he would not change the avenues – believed to be School and Vault Avenues – nor the South Avenue that ran between them. The 3rd Earl had focussed his garden endeavours on planting forest trees on the estate, nearly entirely oak; however, decorative efforts had been few and far between. There was also a terrace across the entrance gates that was planted with large common laurels, flower gardens and beds with roses and honeysuckles, a labyrinth and a 2-acre kitchen garden.

Like the ground-breaking London to Manchester Railway, Barron's first task at Elvaston was also one of engineering – the drainage of the waterlogged grounds. He had to drive a mile and a half before he found a sufficient drop in the ground level to enable any draining activity. The work was long and painstaking but afforded him and his patron time to plan the changes to the estate. Two more different men you would be hard pressed to find. Barron described himself as, 'Highly garnished and seasoned dishes I leave to others, of different tastes; plain, simple, solid food I find best, and therefore keep to it. Having been nourished somewhat in this way, I have grown up as a plain, matter of fact man.' Conversely, the 4th Earl was flamboyant and, as a rich aristocrat, had access to all the delights and variety that life had to offer. And he most certainly enjoyed them. Nonetheless, their relationship was a

Elvaston Castle during the 1800s.

Elvaston Gardens.

positive one. The lively and colourful earl treated Barron 'more as a brother' than a servant and their collaboration was lifelong.

In 1831, after Nottingham Castle was burned down by rioters, Barron repaired the Elvaston estate fire engine, pre-empting damage a few nights later when there was a fire in the castle one evening and in a barn a few nights later. Having got the fire under control before the fire brigade arrived, the story is told that the fire chief knocked Barron off the wall with a blast of water, before being made to apologise by the mayor.

It was based on this trusting and creative partnership that Barron became more and more adventurous with the design of Elvaston. In chapter IV of his book, he starts with the statement, 'The formation of the lake at Elvaston castle, and the rocks on the islands and surrounding banks have been entirely a work of art.' The relatively flat land that Capability Brown found so boring was a challenge for Barron, and one he embraced and expanded with ambition. The plans developed organically too. The lake was started, and part way through excavations the earl decided to increase its size by 50 per cent. The soil from the excavations presented a puzzle at first to Barron, who then struck upon the idea of creating an artificial mound with a wooded walk spiralling to its summit.

Barron then began to extend his creativity, experimenting with rockwork in the garden to create a more natural rolling scenery, breaking up the 'green carpet' of the estate. This activity proved somewhat of a battle between Barron and the earl. The earl liked uniformity and balance, whereas Barron wanted to emulate nature and eventually it was

An example of Barron's rockwork at Elvaston. (T. Liddle)

through Barron's ingenuity and persistence that they found a middle ground. Barron mirrored his first rockwork attempt across the lake with an interpretation of a ruined building. It fitted Elvaston's Gothic feel, and the earl gave him carte blanche to finish what he had started by replicating similar rocky mounds and structures across the gardens. Some of the rockwork reached over 40 feet and Barron endeavoured to make the additions look seamless, planting them with specimens from his palette of trees.

Much of the smaller rockwork still survives today and is loved by visitors of all ages for its interesting shapes, views and picture-perfect spots. Just like Barron desired it upon its public opening in 1851, the variety of sights around the park were designed so that the 'interest of the visitor is continually kept up' – an epithet still true in the twenty-first century.

Barron also developed the kitchen gardens further, extending them from 3 to 8 acres, with five glasshouses and south-facing slopes abundant with greengages, apricots and peaches growing in asphalt to ensure they ripened together with melon and pineapple pits. According to Robert Glendinning, one of few people to see the gardens at Elvaston during the 4th Earl's custodianship (together with the Duke of Wellington), Elvaston boasted the finest apricot walls in Europe.

His first few years at Elvaston did not just represent a significant professional milestone, as it was in this period that he met his first wife, Sarah Allestree (1812–38), and within two years they started a family.

Their daughters Elizabeth (1836–65) and Mary (1838–1901) both outlived their mother, who died aged just twenty-five and was subsequently buried in the graveyard at St Bartholomew's Church on the estate.

In 1841, Barron married for a second time to Elizabeth Ashley. Together they had a son, John (b. 1844), who would later follow in his father's footsteps and become a horticulturalist and would create with his father a family business – Barron & Son.

By the time of his death, Barron had married his third wife – Frances. Having grown up in Norfolk, Frances moved back to her childhood county after her husband passed away. There she named her new home 'Elvaston House'.

The Kitchen Garden at Elvaston. (T. Liddle)

# 5

# Elvaston Matures

*The greatest work of gardening skill, which perhaps one man ever accomplished in one lifetime before, but being kept strictly private, it is scarcely known to exist.*
                                              The Gardener's Chronicle (1849)

Barron, from the start of his studies, had been 'passionately fond' of evergreens and saw a chance to realise a lifelong ambition of his – to plant them on large scale. The 4th Earl, who Barron remarked 'naturally possessed, in extraordinary degree, the organ of colour' agreed to his plan to arrange and plant evergreens in such a design as to 'produce a pictorial effect in the landscape' using 'the great variety of tints' found among them. His experience at the Botanic Gardens, where he had learned to value and enjoy the many varieties, also ensured he knew which of the hardiest species to plant. He struggled at first to convince his employer of some of the lesser-known varieties, but in 1835 he was able to successfully embark on planting a pinetum (a collection pines planted to be enjoyed) with his employer's approval, and this time it was given with significant enthusiasm.

Barron's complex topiary designs at Elvaston.

The varieties of evergreens used across the country in the early part of the nineteenth century were limited. Barron had been lucky to study them at the Botanical Gardens, but further afield he remarked that the scope for the planter was so constricted 'that to these circumstances may be attributed in a measure the almost universal jumble of ordinary-looking trees ... even from the palace to the cottage'. The more common varieties which peppered the landscape included elms, ashes, sycamores, poplars and Barron didn't beat about the bush with the description 'any other rubbish that the nearest provincial nursery may happen to be overstocked with'.

This prevalence of deciduous trees infuriated him and did not help gardeners in maintaining pristine grounds, especially when seven months of the year were spent collecting fallen leaves. Barron loved nearly all trees and was keen his descriptions of the deciduous weren't taken as a rally against nature. He was simply adamant that planting a tree must be done to satisfy a specific objective, not just indiscriminately because they were cheap or easily available. He likened the act to 'planting an orchard of fruit trees, without any regard to the value of the fruit when grown' or wrote, in his inimitable style that when looking in a stable 'you may imagine my disappointment to find only a cow in it ... If I have but one stall in my stable, and keep for only one horse, I think I can manage better than put a donkey in it'.

He wanted trees that would last the whole year and bring with them a 'beauty of form or outline'. The passage in his book describing the options simply sings with his love and passion for evergreens:

The Garden of Fair Star.

The formal Araucaria, and fastigiate Junipers, to the wild grandeur of the pine, and even to the delicate, graceful and flowing habits of the Japanese Cedar, Funeral Cypress (Cupressus funebris), Deodar Cedar (Cedrus Deodara), and Hemlock Spruce (Tsuga canadensis), – or size, from the gigantic Lambert and Bentham Pines, the Sequoia sempervirens, and Douglas Fir, towering their lofty heads a hundred feet above the pride of British forests ... to the diminutive dimensions of the Siberian pine, Hudson, Clanbrassil, and pigmy firs; – or colour, from the deep sombre tone of the common Irish yews, the glaucous and silvery in the Sabine Pine, P. nivea and Deodar Cedar, and to the warm and yellow greens in Pinus insignis, and other even to the rich yellow of the Golden Yew.

It was with this absolute dedication that he set about the planting of the evergreen pinetum. Unfortunately, it came with difficulties. The 4th Earl had little horticultural knowledge himself and, unlike many of the aristocratic class with large estates, was not a member of the Horticultural Society. This made it difficult for Barron to source the variety of specimens he wanted until they became commercially available. Often specimens were found as individuals, which led to Barron starting to propagate multiples to satisfy their needs. This would prove a lifelong interest and passion for Barron, who would eventually open his own nursery after his career at Elvaston.

The landscape gardening practices developed and employed at Elvaston, which Barron regarded as 'one vast pinetum artistically treated', especially the hugely ambitious tree-transplanting, propagating and grafting, transformed a largely featureless site into one of the most celebrated gardens in Europe and North America.

Elliott, Watkins and Daniels, Garden History,<br>Vol. 35, (2007), published by The Gardens Trust.

An Example of Barron's tree planning at Elvaston. (T. Liddle)

# 6

# The British Winter Garden

*The best planter of evergreens in the world!*
*The Farmer's Gazette* (1851)

Barron's love of evergreens permeates the planting at Elvaston – it is as evident today as it was in the 1800s when he first laid out the gardens. His passion for evergreens was lifelong, and he believed them the best and most appropriate palette for the British climate.

His book, first published in 1852, was unambiguously titled *The British Winter Garden: Being a Practical Treatise on Evergreens; Showing Their General Utility in the Formation of Garden and Landscape Scenery*. It covers the history of his planting at Elvaston and details the benefits of evergreens, their varieties and propagation.

Barron's desire to write the book came part way through a career in which he was able to explore planting on a large scale. His time at Elvaston had 'enabled me to make observations and fully test what can be accomplished in a given time, and to offer suggestions, which if carried out, will in a short time materially alter the value and aspect of our country'. He humbly, but truly, believed his expertise would change the face of British gardening – something that can be seen in our review of his many works later in this book.

His advice to gardeners is direct: 'Whoever can afford a piece of ground for the growth of an ornamental tree, should contrive to have a tree which will prove an ornament, both by its innate beauty and vigorous, development after it is planted.' Rather than leaving this as a simple statement, Barron goes to great lengths in *The British Winter Garden* to ensure this style of gardening is accessible and applicable to those who wish to engage with him.

His tried and tested methods are outlined in various chapters of his book. As well as explaining his approach to, and the outcome of his work at Elvaston, he works through the principles any aspiring gardener needs to plant the quintessential British winter garden. He covers the transplanting of large trees – made famous by his transplanting machine – to enable a mature look quickly. He discusses at length the 'evils arising from the pot-culture of such Plants as should ultimately become Trees', resulting in damaged roots and stunted growth. This is a key issue for Barron and one that was arising from the growth in demand of trees from nurseries. The need to produce trees quickly for the public resulted in horticultural methods akin to mass production. The public buying

Belper Cemetery. (T. Liddle)

ready-grown and ready-to-plant trees and were either, blind to, or deliberately ignored, the reduced quality and higher potential for failure this method resulted in. Barron knew the pitfalls and extolled the virtues of raising plants from cuttings, grafts, and layers. While he acknowledged that this a more complex and time-consuming way to achieve the desired outcome, ever the consummate gardener, he saw no acceptable alternative.

Barron did not just stop at the buying and cultivating of trees, he considered their location, their arrangement and highlights a selection of the most valuable trees and shrubs available to the reader of the time. He expands later in the book to cover evergreen hedgerows and their utility for sheltering cattle as part of a wider working landscape.

Such is the climate of the British Isles that the use of evergreens is a natural and logical choice for the savvy gardener. Many of the evergreen trees introduced during the 1800s came from countries in which the ability to survive the cold was a necessity. This made them perfect for Britain and meant that Barron could focus on the more sensitive varieties which needed protection through the variability of our climate. Barron talks about his observations on the types of soil suited to each plant after lengthy trials, enabling trees to weather spring frosts and summer heat waves alike.

The recent introduction of the evergreens outlined in his book did, however, pose logistical challenges at the time. Their scarcity meant that it was harder to plant at scale, but this did not stop him and or his encouragement to others. He firmly believed in the illustrious nature of his contemporaries: 'Once it is known that these things will be wanted, the enterprise of British nurserymen will soon supply them plentifully at a reasonable rate.'

The East Avenue at Elvaston.

Barron became one of these nurserymen. He had long grown the varieties he needed at Elvaston on the estate, and later became one of the most renowned commercial nurserymen in the country. He worked tirelessly with his son to bring the evergreen varieties he so extolled to the masses.

His arguments against deciduous varieties are strong and chief among them is the 'continued litter of decayed leaves during that period of the year when our gardens are expected to look their best; an assemblage of leafless stems without either beauty of form or outline'. Barron approached garden design and planting from both an aesthetic and a practical perspective and deciduous trees didn't support the best outcome in either of these areas. As a designer with an exacting client, he wanted to produce the best, most interesting and diverse designs, which gave year-round joy and value. As a head gardener responsible for maintenance, he wanted a landscape that was straightforward to manage; gardens on this scale needed to be sustainable and not require armies of attendants to keep pristine.

This is where, in Barron's mind, the evergreen family of trees comes into its own. 'I should infinitely prefer planting it *exclusively* with *evergreens* wherever beauty or shelter was my motive for planting. By this means, I should secure the enjoyment of my grounds (in these points at least) for twelve months in the year instead of only a few months in summer.' The varieties and shades give 'a play of light and shade peculiar to themselves' which he found matched nowhere else.

You would be forgiven for thinking that perhaps, with such a strong focus on evergreens, Barron disliked and saw no value in deciduous varieties; however, this was not the case.

Barron's rockwork and planting come together at Elvaston's lake. (T. Liddle)

They had their place and 'should be so disposed of as to secure the greatest possible advantage from their beauties and uses; but should never be allowed to occupy the place of such as will be both more useful and ornamental in a shorter space of time'.

Barron's expertise and experience were of national renown, and he was called as the first witness to a parliamentary select committee on the establishment of renewable forests.

Perhaps Barron's approach is perhaps best summed up in his own words: 'Not any tree should be planted without an object in view to justify the act.' Methodical, precise and exacting in his demands, the choice of plant for any garden is still just as important. His advice is still just as relevant today and, while overlooked in favour of more celebrated gardeners, *The British Winter Garden* remains one of the most comprehensive, expressive and passionate books on how to make the most of British landscape for year-round enjoyment.

Nottingham Road Cemetery. (T. Liddle)

# 7

# The Tree-transplanting Machine

(Chapter collaboration with The Gardens Trust)

Undoubtably, one of Barron's most remarkable achievements is his tree-transplanting machine. Its application at Elvaston and Royal Botanic Gardens, Kew represents perhaps the best-documented use of the technology in action; however, attempts to transplant trees proceeded Barron. We know little of the early tree-transplanting practices. At Versailles, Le Notre was reported to have used a basic apparatus to dig a trench around the roots and the prize them from the ground. The next summary appears in Charles McIntosh's *The Book of the Garden.*

Charles McIntosh had been a head gardener at several significant estates such as Claremont under the patronage of Leopold, King of the Belgians and looked after Dalkeith under the Duke of Buccleugh. He was, importantly, a friend of John Claudius Loudon, another esteemed and well-travelled Scottish gardener who had investigated and developed early tree-transplanting techniques. Along with his contemporary John Lindley, Loudon was at the forefront of horticultural innovation.

*The Book of the Garden* was McIntosh's sixth book, and he outlines the progression from Le Notre. The great Capability Brown is noted as having utilised a 'janker', i.e. a basic device used for transporting logs, and that it was a 'defective' piece of machinery. Sir Henry Steuart writes in *The Planter's Guide* (1828) that it consisted of a strong pole with wheels at its extremities that could be pivoted to remove and replant trees. Steuart reports that it could sometimes go awry, with some of the trees being transported overwhelming the cords that held them in place. This had the effect of the tree breaking loose and effectively propelling the machines operators into the air, landing 'many yards' distance away'. Thankfully, McIntosh notes that transplanting technology had improved and those 'constructed upon correct principles' had become more prevalent and outlines a useful summary.

Matthias Saul is the first listed with his 'Mr Saul's tree and shrub lifting machine'. It was a simple contraption and 'exceedingly well calculated for the removal of subjects under half a ton in weight.' Mr Saul was a Lancastrian gardener and nurseryman who devised a whole series of other 'interesting' gadgets and appears in the gardening press from the 1820s until his death in the 1860s.

1468. *Tree-transplanting machines* of two or more species have been invented. The pole and wheels (*fig.* 210.) is for general purposes the best of any of them. It consists of a long beam or pole, attached to an axle and wheels. The tree being prepared for removal, and the pole placed in a vertical position against it, the stem or trunk is attached to it by ropes; thus attached, they are brought into a horizontal position, by men or horses, with the ball of earth attached to the tree. Horses may then be yoked to the axle at the opposite end of the pole, or root end of the tree, with or without the aid of another axle, and the tree drawn to any distance and planted. In favorable climates, and when a little extra expence is no object, astonishing effects may be produced by removing large trees; and no machine is better adapted for aiding in the labor than this simple union of the pole and cart-axle.

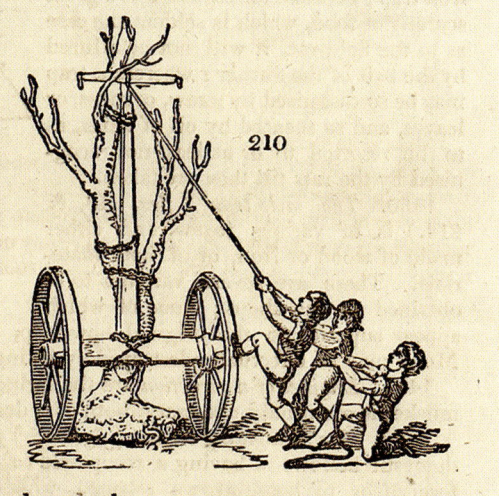

1469. *The German devil* is a frame of timber, with a cylinder moved by a combination of wheels, and a winch, as in raising clay or earth from pits or mines by manual labor. But instead of the bucket of clay, three hooks are attached to the end of the lifting rope, and these are fastened to the roots. (See *Hunter's Evelyn's Sylva*.)

1470. *The hydrostatic press* (*fig.* 211.) may be applied to the same purpose as the *German devil*, with incomparably greater effect. The only difficulty is in finding a proper and convenient fulcrum; that done, this engine will root out the largest trees. It is successfully employed by engineers in drawing piles, gate-posts, raising stones, &c. (See *Nicholson's Arch. Dict.* art. *Hydrostatic Press*.)

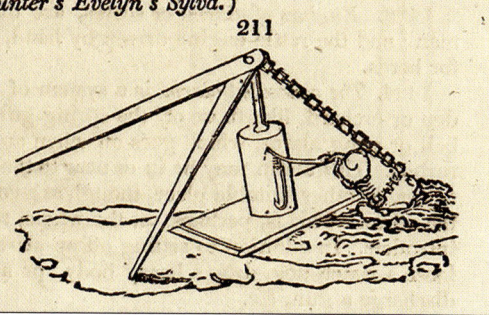

1471. *The garden-seed separater* is a small

From Loudon's *Encyclopedia of Gardening*.

Its use was relatively straightforward. Once a trench had been dug around the tree, one part of the frame (in the foreground of the picture) was dropped into the trench on each side. Iron rods were then pushed through the holes on one side, then through the soil and tree roots, and then finally into the holes on the other part of the frame. Once in place, they were locked and held the rootball and its earth in place. The frame and the tree were then pulled out, using horsepower, and taken off to its new home where the process was reversed. Mr Saul also suggests that the removal, especially for more delicate plants, could be done in stages – the rods were inserted and left for several months so the roots grow round them before lifting.

The principle of allowing the tree to partially regrow before removing it underpinned the next of the inventions described by McIntosh. This was Standish and Noble's crate, devised by two nurserymen to help prepare trees in their nursery grounds for moving after they had been sold. These 'enterprising cultivators' suggested growing trees 'for a season or two in skeleton boxes or cradles' made from elm slats with lots of gaps for the roots to grow through. When planted out in their final position the trees 'need not be

37

Mature trees at Elvaston. (T. Liddle)

removed from their crates, as they will be quite rotten before the roots are of sufficient size to be obstructed by them'.

The next iteration was found in Edinburgh with Barron's mentor, James McNab of the Royal Botanic Gardens. He is reported as having had great success and 'for years attended the removal of trees, both deciduous and evergreen'. McNab devised a whole series of devices for lifting plants of different sizes, which were 'simple and easily constructed'. The one used for moving large trees worked on a series of ratchets, winches and rollers, reducing the physical effort involved. No doubt, Barron would have been learned a great deal from McNab during his time there, enabling him to perfect his own designs at Elvaston.

McIntosh himself could attest to efficacity of McNab's machine. He used one at Dalkeith 'made of the very best materials and workmanship, [which] cost £25. It has been in use 11 years and has not required the least repair. It is capable of removing trees to any distance. He notes that 'we once brought a tree to Dalkeith from the neighbourhood of Glasgow, 46 miles, with one horse, with the greatest ease'.

Barron's machine directly follows in the footsteps of Capability Brown, Mr Saul and McNab. Barron had already published an account of his machine and methods in his own book *The British Winter Garden*. It was constructed out of huge oak beams – the longest

Illustration of a tree transplanter from McNab's scrapbooks held at Royal Botanic Gardens Edinburgh. (© Royal Botanic Gardens Edinburgh)

21 feet long – and iron rods and mounted on a pair of massive broad-tired wheels. Despite this it was easy to disassemble, manoeuvre into position and reassemble, and it is very clever. It involved an advance on earlier systems by getting a cradle right under the rootball.

Barron also built a second machine for moving smaller trees of up to a couple of tons in weight. It worked by a system of cables and windlasses, which McIntosh describes:

> In weighing an anchor, the cable is wound round and becomes at each turn more and more shortened, [so] the platform and rootball upon it are elevated until sufficiently clear of the ground. By these simple mechanical appliances, the great transplanting operations carried on at Elvaston during the last twenty-five years have been executed.

Many of the trees at Elvaston had been moved as mature specimens by Barron's machinery. They included the large yew, which was approaching a hundred years old and had been moved 25 miles. Probably even more famous was his successful transplantation of the 800-year-old Buckland Yew at St Andrew's Church, Buckland, near Dover, to create room for an extension to the Church.

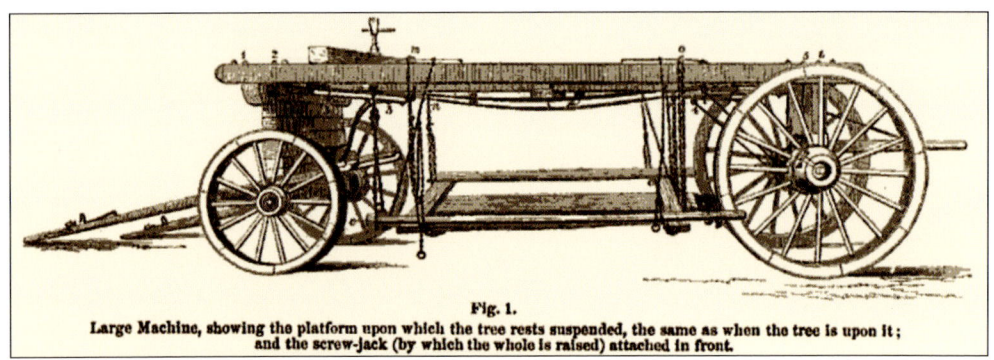

**Fig. 1.**
Large Machine, showing the platform upon which the tree rests suspended, the same as when the tree is upon it; and the screw-jack (by which the whole is raised) attached in front.

*Above*: Barron's large transplanter.

*Left*: The removal of the Buckland Yew. (*Gardeners Chronicle*, April 1891)

We also know that Barron's tree-transplanting machine was used at Royal Botanic Gardens, Kew, in the 1880s and later. The machine still survives, the only known specimen, and is housed at Kew. In the year 2000 it was used to plant mature trees along a new avenue at Kew in celebration of the millennium. After 140 years, it is still in remarkable condition and still able to transplant trees in the manner designed by Barron.

The surviving Barron transplanter at Kew Gardens. (T. Liddle)

*Right*: The surviving Barron transplanter at Kew Gardens. (T. Liddle)

*Below*: Tree transplanting – preparing tree base and root system for removal at Royal Botanic Gardens Edinburgh, late 1880s. (© Royal Botanic Gardens Edinburgh)

# 8

# Elvaston Opens to The Public

*The opening to the public in 1851 of Elvaston Castle, Derbyshire, with its multiple avenues of conifers, each row taller than the one in front, had an immense impact on the gardening world. The pinetum, which was described by Glendinning as an example of the sublime in gardening, consisted of two symmetrical sections, each bisected by an identical quadruple avenue, with one section devoted to pines and the other to spruces and firs.*

Elliott, *Victorian Gardens*, pp. 83–87

The south front of Elvaston Castle with gazebo.

With their opening, the gardens attracted much attention. There was dislike of the artifice used to create the gardens – the use of rockwork, the transplanting of mature trees and the topiary of the plants that achieved a precise look. There were also concerns that the meticulous pruning and snipping would be detrimental to the health of the trees. Happily, Barron's work at Elvaston proved the critics wrong. The gardens and plants thrived and Barron himself knew that the criticism was most likely born out of fear of the new and different. The overwhelming reception, however, was positive, with one observer stating at the time: 'When we contemplate, for a moment, that this is entirely a work of art, and consider the tens of thousands of tons of rocks all brought from a great distance, employed in its formation, we are left to conclude that it has not only no rival as a work of art, but there is nothing at all approaching it, in any garden in this country.' (Robert Glendenning, quoted by Elliott in 1986, pp. 95)

*Right and below*: The Italian Garden at Elvaston.

The old carriage drive at Elvaston. (T. Liddle)

Compared to some of his contemporaries, Barron was relatively late in his career when he started to undertake private landscape gardening, some of which we explore in this book. It is likely that it was the encouragement of the 5th Earl of Harrington that persuaded Barron to undertake designs for commissions for private gardens, public parks and cemeteries. A deeply religious man, this work was important to Barron who perceived it as a way to further his own moral and spiritual objectives, as well as to promote the use of evergreen planting. He believed that the development of evergreen gardens would encourage year-round opportunities for recreation in urban areas.

# 9

# Messrs Barron & Son, Borrowash

After his last patron, Leicester Stanhope, the 5th Earl of Harrington, died in 1862, Barron purchased a substantial piece of land in the neighbouring village of Borrowash – 40 acres all told. He purchased the land from the earls of Harrington, an act that tied his fate to theirs until the end of his life.

The stock from the estate nursery, which he had developed from scratch and had also latterly rented as a business in its own right, was moved from the estate to Borrowash in around 1862. It was here, after a long and successful career, William Barron retired in 1865. Not satisfied with a leisurely retirement, he formally opened up Elvaston Nursery with his son, John. In 1874, he also opened a shop at No. 16 Market Street, Nottingham, to sell plants, seeds and bouquets of flowers. Listed in the 1881 census at the age of seventy-five, Barron is noted as a 'Landscape Gardner and Nursery-man' and his operations extended to 30 acres with forty-one staff.

Barron's nursery was vibrant and successful, opening at 6 a.m. every morning. The nursery supplied trees, seeds and wreaths all over the country. There were numerous people involved in the creation and dispatch of the latter, including a full-time carter who drove the goods to the station for their nationwide journeys. The nursery was laid out methodically, much like Barron's approach to his commissions. Glasshouses and lily ponds, reminiscent of the botanical gardens, were also a feature – perhaps inspired by a lifelong desire to recreate the environments where he had spent some of his formative years.

He presumably drew inspiration from the challenges he faced when he tried to find specimens to fill the pinetum. Consequently, he made it his mission to ensure that the plant offering at Borrowash contained every new plant introduced into the country over the preceding twenty years. He enlisted agents across the globe to send packages of the best quality seeds to him, which the nursery staff then propagated. The international connection continued with commissions received across Europe, and a select number of international staff came to work with and learn from Barron at the nursery. Barron even sent his son John to Silesia and the Netherlands for part of his education, being taught by Prince Herman Puckler-Muskau and renowned landscape gardener Eduard Petzgold. In return, Petzgold's son worked at the nursery for several years as well.

Photo of William Barron's tree transplanter planting an *Auricaria* at Royal Botanic Gardens Edinburgh, taken by Adam Dewar Richardson in the late 1880s. (© Royal Botanic Gardens Edinburgh)

Critical to the success of any nursery during this period was the network of connections that it could make. Having successful relationships with gardeners, collectors, journalists and writers who could both buy the plants and create publicity was fundamental. Barron's reputation secured these connections, and the nursery was able to thrive. However, the nursery still lived with the competitive pressure of ensuring that it had the best selection of quality plants, and that there were always new and novel specimens for sale. Unusual varieties helped nurseries corner the market and attract wealthy collectors, and in turn enable them to sponsor adventurous explorers going in search of the new.

While exciting for the consumer, the challenge lay in ensuring that the specimens, many derived from significantly different climates, survived in the British ground and its weather. Barron's experience as the Head Gardener at Elvaston and at the Edinburgh Botanical Gardens gave him direct and invaluable experience of the realities of the luxury of longer-term estate and academic propagation, versus quick growth for commercial use. He commented on the widespread use of pots to grow specimens, complaining that the damage to the roots of the plants was irreversible and not rectified even when planted in the ground. An entire chapter of *The British Winter Garden* is dedicated to this ('The Evils of Pot Culture').

The nursery became known for its wealth of coniferae, which was emphasised in their catalogue, in addition to a wide range of shrubs and plants – from the utilitarian to the ornamental. Its fame grew as it exhibited at horticultural shows across the country, often

Tree transplanting – getting ready for haulage at Royal Botanic Garden Edinburgh. Taken by David Sydney Fish, late 1880s. (© Royal Botanic Gardens Edinburgh)

winning first prize. Barron and the Elvaston Nursery are perhaps most famous for the development of the Elvaston yew (*elvastonensis aurea*) – a golden 'bright orange colour, and unlike all other or silver yews, is not variegated but a self-colour; it is by far the most brilliant in winter'. It gained fame and was even described in George Gordon's *Pinetum* book (Barron had helped develop the actual pinetum) as 'by far the most brilliant of the golden varieties in the wintertime'.

The nursery also stocked many other beautifully coloured varieties, such as the *Taxus baccata variegate baroni* – Barron's variegated yew. Fitting that a man who had spent his life creatively using and nurturing evergreens would have a variety of one his undoubtable favourites named after him.

The selection was contained in the multiple catalogues produced by the nursery:

- *Catalogue of Coniferae, Ornamental Plants, Forest Trees, etc.*
- *Catalogue of Coniferae, Hardy Ornamental Trees & Shrubs, Hardy Evergreens, Forest Trees, etc* (circa 1900)
- *Catalogue of Fruit Trees, Roses, Herbaceous & Alpine Plants, Greenhouse & Bedding Plants, etc.*
- *Descriptive Catalogue of Coniferae and Other Ornamental Plants 1882/83*
- *Catalogue of Forest Trees, Cover Plants, etc.* (1889/90, 1894/95, 1895/96, 1896/97, 1897/98; Contains information on W. Barron & Son's tree-transplanting machines invented in 1831)

Advert of unknown date (possibly 1930s).

In 1881, Barron moved into Elvaston House on The Turnpike in Borrowash. His son John was living at Rock House, on Brook Road, close by. Both houses are still standing today.

Elvaston House was amid the operations of the nursery and Barron had large bay windows added so he could survey the work of his staff. It was in this house, at the age of eighty-five on 11 April 1891, that Barron passed away. He was buried close to his beloved Elvaston gardens, home, and nursery.

Even after Barron's death, The Elvaston Nursery continued from strength to strength, supposedly becoming the biggest in Britain, growing over a hundred thousand roses every year for discerning customers, as well as continuing to develop its range of conifers. John Barron gained commissions locally in Derbyshire (Broomfield Hall, Risley Hall, and Matlock Bath Pavilion) and abroad, including a city park in Tunis. In true Barron style, the nursery itself was well designed and attracted many visitors annually just to see its landscaping, pools and fountains.

John later ran the nursery for a further twenty years and was a farmer in his own right. A founder member of the Nottinghamshire Agricultural Society (1879), he bred pedigree breeds such as shire horses, shorthorn cattle and white Yorkshire pigs. The latter led him to the National Pig Breeders Association where he was a respected council member and competition judge. John died following a seizure, which started while out hunting with the Earl of Harrington's hounds.

Elvaston House, Borrowash. (T. Liddle)

Barron's daughter Amy was also an agricultural force in her own right. Born in 1862, Amy was fascinated by the world of dairy, training at Lord Vernon's Dairy School in Sudbury and lecturing at diary institutes around the country. At the Glasgow Exhibition of 1888, she is noted as having demonstrated butter churning and ventured far further afield to Jamaica for its Great Exhibition in 1891. Her obituary, following her death in 1892, commemorates her as a 'highly educated and accomplished ... popular dairy expert.' Not to let the side down, his other daughter Margaret married the Professor of Agriculture at Bangor University, Thomas Winter.

Barron's continued to be a key employer for the local village residents. One resident, Eric Collyer, went on to a found his own nursery (Collyer's), which is still operating in Borrowash today.

The arrival of the First World War signalled the beginning of the end for Barron and Company. Now a limited company, swathes of its land were sold off at a profit, decreasing the size of the operations. Then, after the Second World War, it slowly became unviable. Ninety-seven years after it first opened, the Elvaston Nursery closed its doors for the last time in 1959, twenty years after the family ceased their involvement with the business.

# 10

# The Gardens at Elvaston Today

Although somewhat faded in its grandeur, Barron would still recognise much of his work at Elvaston Castle. After Barron left, Elvaston remained the home of the earls of Harrington for several more decades, the family leaving in the 1930s for their home in Ireland. There is not much written or recorded about the gardens after the 5th Earl, but it appears that some areas were much simplified (a process that had begun under Barron himself with a much reduced gardening staff). Over a number of years, the Alhambra Garden became a Badminton Court, the North Lawns became a croquet field and while the gardens retained their original designs elsewhere, it was inevitable that less attention was paid to maintaining Barron's originally tightly trimmed topiary.

During the 1940s a teacher training college occupied the house (after being evacuated from Derby during the war) and images from this time clearly show areas such as the Garden of Fairstar looking much overgrown, but still fully recognisable. When the college left in 1949 the house remained empty until 1963, with just the Harington's estate manager living on the site in the thatched cottage. In 1963, the contents of the house were sold at auction and the property purchased by Needlers Development Corporation. The firm, a precursor to Tarmac, had planned to quarry the estate for gravel but, having failed to get permission, sold the property to Derbyshire County Council and Derby City Corporation in 1969. Some scenes from the film *Women in Love* were filmed at Elvaston in 1969, including a famous naked fight scene in the Gothic Hall and two locations in the gardens – the Golden Gates and the Alhambra Temple (the decorative balcony was used despite never having had any access, or even a floor).

In 1970, Elvaston Castle Country Park opened to the public. It was the first country park to be registered under the Countryside Act of 1969 and the second to be opened to the public. It soon became clear to the new owners that there was a significant backlog of repairs and maintenance required to both the house and gardens. The water tower and a range of service buildings were demolished (thus removing the gravity fed water supply for Barron's fountains) and the Garden of Fairstar was replaced by a much simpler, though still complex, box hedge parterre. While Elvaston has certainly seen its fair share of theft and vandalism over the years it is unclear what the family may have removed, what was sold and what may have been stolen. However, there are certainly less statues, ironwork and architectural features than had existed in the previous century.

Nonetheless, Elvaston retains the spirit of William Barron and the 3rd, 4th and 5th earls who much admired his work. Barron's legacy lives on at Elvaston: his rockwork, pinetum, sunken garden, Italian Garden and even the Garden of Fairstar are still clearly identifiable. So too is the Alhambra Garden, the Azalea Garden, the lakes and water features and the avenues he created. Many of Elvaston's trees are those that Barron transplanted as mature specimens (creating some arboricultural challenges for the property today). Elvaston is still home to the varieties of Yew that Barron created and much of its appearance is a direct result of botanical propagation techniques Barron perfected in the once extensive glasshouses – few of which survive and all of which need restoration. Had Elvaston's future been mapped out differently in the early part of the twentieth century the gardens would today be one of the most significant gardens in the country. As it is they remain the best example of Barron's work on a private estate, and they are the origin of the techniques and methods he mastered that allowed Messrs Barron & Sons such phenomenal success as gardeners, landscapers and nurserymen through to the 1930s. As hero gardeners go, and as has been noted in other recent gardening research, William Barron possibly made some of the most significant contributions to landscape design, horticultural practice and public gardening of any of his contemporaries, many of whom have found fame only because their masterworks survived in much better condition.

The Gardener's Bothy at Elvaston. (T. Liddle)

# 11

# Barron's Legacy

*To plant a garden is to believe in tomorrow.*

Audrey Hepburn

It's hard to imagine, when you review the breadth and depth of his work, that there was a more prolific or impactful gardener in Victorian England than William Barron. The company William Barron established in Borrowash in the 1860s undertook hundreds of gardening and transplanting commissions throughout Britain and Europe. Barron had

William Barron's blue plaque at Elvaston House, Borrowash. (T. Liddle)

developed the first modern commercial nursery operation and created a business which was to continue in to the 1930s.

At Elvaston he created a stunning and nationally significant garden where Brown famously said there was 'no capability'. His skills and knowledge learnt in Scotland, and his vision for creativity, provided strong foundations for the creation of a paradise for two lovers – the 4th Earl of Harrington and Maria Foote.

While at Elvaston he created a particularly efficient tree-transplantation machine, which was to provide the basis for much of his commercial work. During this time, he also wrote *The British Winter Garden* – a book that has influenced generations of gardeners and continues to influence garden design today.

His development of different varieties of yew and coniferae in many different colours addressed the criticism of evergreen trees being too dull – thus overcoming the principal objection to their overuse and instead creating demand for evergreen features.

His work also featured in two publications, *Pinetum Britannicum* and *Gordon's Pinetum*. It would be impossible to know how many gardens were influenced by these books, but there is no doubt that it was Barron's work that informed the design of the pinetum at Biddulph where many trees were planted on mounds and golden yew hedges used.

His work at Elvaston raised the profile and popularity of evergreens, yews and topiary. It should be noted that the style of tree planting that is seen at Elvaston (where the rootball is covered with earth to create a mound, so that as the tree matures the roots become visible and of botanical interest) was a technique developed by John Claudius Loudon but probably popularised by Barron.

In 1887, Barron was invited to Parliament to establish forestry and arboriculture as recognised professions in Victorian society and to advise on the development of renewable forests.

During his lifetime, Barron changed the way that gardeners and nurserymen propagated and maintained tree and shrub specimens. He developed techniques and methods that rapidly transformed open spaces into mature gardens, enabling the commercial supply of many new varieties of plants. His legacy should not be forgotten, as his role in the development of modern British gardening is perhaps greater than his contemporaries. His work lives on at Elvaston and in numerous public parks around Britain – parks that are visited by hundreds of thousands of people every year.

A man passionate about green spaces and their impact on their visitors, Barron's work continues to touch the lives of families and individuals seeking space, enjoyment and relaxation in an increasingly urban society. His grandest and most complete work – Elvaston – has also now become a public park for generations more to enjoy and be inspired by. We can think of no better legacy for a man who gave so much to shaping the British gardens we know and love.

'At all events, however far the desired end may be accomplished, I shall have the satisfaction of knowing I have made an attempt at improvement for the benefit of others.'

William Barron, *The British Winter Garden*

Rockwork at Elvaston. (T. Liddle)

# 12
# Private Commissions

Tree transplanting – operation on a walnut tree at Royal Botanic Gardens Edinburgh. Taken by Adam Dewar Richardson, 14 November 1894. (© Royal Botanic Gardens Edinburgh)

Aqualate Hall.

## Aqualate Hall, Staffordshire

Although there are records that Baron worked at Aqualate Hall, a property that dates back to the early-seventeenth century, little information is available. The hall was rebuilt by John Nash in the late eighteenth century in the Gothic style, and this is likely the time that the grounds were designed by William Barron. In 1910, the hall was destroyed by fire and replaced by a new building between 1927 and 1930, which was designed by William Douglas Caroe.

While there is little information available about William Barron's work, the postcard shown here does feature conifer and flower beds, which reflect Barron's designs elsewhere.

## Aston Lodge, Aston-on-Trent

Aston Lodge was purchased by Reginald Boden (1872–1952) of the Derby lace-manufacturing dynasty. He was third son of Henry Boden of the Friary and his mother, who came to live with him, was a Holden of Aston Hall.

Among the works he undertook to the house, he commissioned a modern stable block, added a water tower, added a second bow window on the garden front, rebuilt the ground floor on the entrance front to include a new door and deeper reception rooms, and raised the ballroom wing to make the façade symmetrical. This work was likely carried out by Alexander MacPherson of Derby.

Aston Lodge.

He also commissioned William Barron & Sons to relandscape the grounds, installing gates by Robert Bakewell (and the associated gate piers), which had once stood on the river front of Rivett House in Tenant Street, Derby.

## Betley Court, Staffordshire

Work commenced on the gardens, parkland and house at Betley in 1730, and the landscape was subsequently redesigned by William Emes in 1783. In 1865, the property passed into the ownership of Thomas Fletcher-Twemlow, who employed the county surveyor Robert Griffiths to undertake work including the construction of a dam to create a small pool and install a hydraulic ram. It was at this time that William Barron was engaged to design formal gardens and parterres to the south of the house. Continual changes to the gardens since this time mean there are few remnants of Barron's work – the terraces are the only significant survivor.

Barron's plan of 1866 shows a French-style parterre with a central ornamental feature. A large cedar or Lebanon is one of the most impressive specimen trees and is thought to have been brought to the estate by William Barron in 1865. The picture of Betley Court in this chapter is from 1870. There were two kitchen gardens – one by Emes and one by Barron, both with high red-brick walls with swept coping, edged with blue brick and with ball finials. These were lost beneath a housing development with only a part of a chimney remaining from Barron's former heated glasshouses. The register of parks and gardens notes that Betley does not have sufficient remains of the work by the two most influential contributors – Emes and Barron – to be worthy of registration.

Barron's Yew Tree Screen is seen in this garden plan (highlighted) and in the left of the photograph below. The more recent image shows steps through the screen.

Betley Court. (Sue Hurrell/Betley Court)

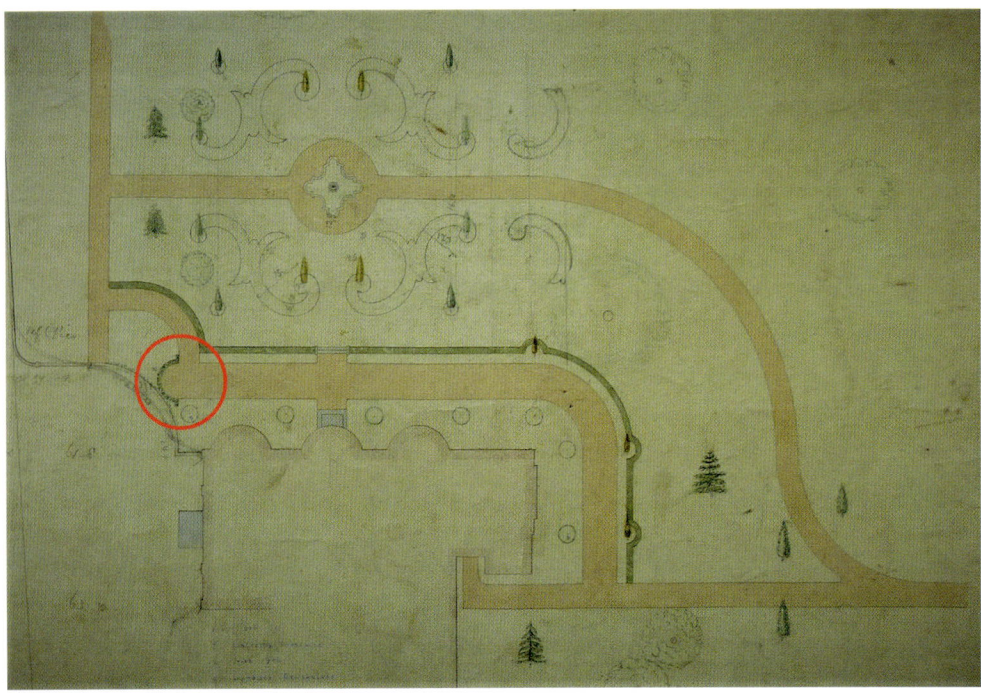

*Opposite below and right*: Barron's yew tree screen. (Sue Hurrell/Betley Court)

*Below*: Cedar of Lebanon, planted by William Barron. (Sue Hurrell/Betley Court)

# Bretby Park, Derbyshire

Little remains of the original gardens at Bretby Hall. The original gardens had been laid out between 1684 and 1702 by Philip Stanhope, the 2nd Earl of Chesterfield (and related to the Stanhopes at Elvaston). Earthworks around the site hint at these early designs.

In 1924, William Barron Ltd was called in by J. D. Wragg to relandscape the gardens, but two years later the property was sold to Derbyshire County Council. It operated as a sanatorium to treat children and later became an orthopaedic centre before closing in 1997. The house has since been converted to residential apartments.

# Broomfield College, Derbyshire
(Courtesy of Adrienne McStocker)

In 1870, Charles Schwind (1822–99) purchased farmland a few miles north-east of Derby for £8,000 to create a country estate. His family originally came from Baden in Prussia and made their money in the South American trade via Liverpool; however, Charles Schwind's business interests expanded into Derbyshire. He became chairman of the Handiside Ironworks, a director of the Crompton & Evans Union Bank and a Justice of the Peace. The Schwind story is that of a successful Victorian merchant family and reflects the rise of the middle class in nineteenth-century England.

To build a house suitable for his large family and commensurate with his social standing within the area, Charles Schwind turned to the architect Frederic Josias Robinson (1833–92) of Derby. The house was constructed in 1873 at a cost of £6,449.

William Barron & Son were engaged to design the gardens and pleasure grounds for the house. By the time the landscaping was undertaken in 1870 to 1873, the firm enjoyed a national and international reputation. There was also a Prussian connection. Barron had a business relationship with Prince Hermann Pückler-Muskau and Edward Petzold, the leading European landscape designers and arboriculturists. His son, John Barron, was partly educated at the Muskau estates in Prussia and the Netherlands and there was a flow of personnel between the two businesses for a number of years.

Barron's plans for the garden and pleasure grounds at Broomfield Hall have not been located, but Ordnance Survey maps from 1880 onwards provide a clear outline. Subsequent maps demonstrate that no major alterations were made to the gardens and pleasure grounds up to the sale to Derbyshire County Council in 1947 when it became the Derbyshire Farm Institute.

The original glasshouses and conservatory no longer remain but their locations are easily discernible on the site today. The underground boiler for the large glasshouse on the northern perimeter of the walled garden still exists, as do a number of original utility buildings, including a root store, a fruit store and a gardeners' bothy. The productive walled garden underwent planting changes post-1947 and today contains demonstration gardens constructed by horticultural students.

The front of the Broomfield Hall surrounded by trees and shrubs. (Derby College)

The 1880 Ordnance Survey map shows an extensive mature treescape on what had been farmland prior to 1870. Barron may have brought in mature specimens to provide instant effect as he had done elsewhere. The original design structure and some of the planting in the ornamental gardens and pleasure grounds remain including the western shelter belt, the sunken garden, and the circular lawn with its quadrant beds. Historically, this was known as Barron's rose garden and is now planted sympathetically with tropical plants. Barron formed grassy banks around the house, supporting the terrace with stone steps down to the lower gardens. On the southern elevation of the house there is a large lawn incurved with trees and shrubs. The southern view extends beyond the ha-ha to the ridge, which hid the city of Derby from view in the nineteenth century.

As the grounds move around to the eastern elevation and the main entrance to the house, the planting framed the house, surrounded a large lawn and lined the driveway.

After 1947 the carpark at the front of the house was extended and the trees on the mound to the right side of the entrance were removed and replaced with what is now known as the 'All Seasons Garden'. The woodland, with its meandering paths on both sides of the driveway, remains and now contains a winter garden. It is still possible to see many fine examples of trees similar to those employed by Barron at Elvaston Castle and Belper Cemetery such as deodar cedar, *Fagus sylvatica f. purpurea, Taxus baccata, Araucaria araucana* and *Thuja plicata*.

Broomfield Hall contains one of the more complete examples of gardens and pleasure grounds designed by Barron for smaller country estates. It provides a useful companion piece to the nearby gardens at Elvaston Castle. Broomfield Hall is now the land-based campus for Derby College and the gardens are open to the public.

*Above and below*: The south front at Broomfield Hall and the southern vista. (Adrienne McStocker, 2021)

# Foremark Hall, Derbyshire

During 1910 and 1911 William Barron & Sons laid out new gardens to the south of the hall. This is one of the few projects undertaken by the firm where extensive records are still available, providing an insight into the firm's work and illustrate a long-term relationship between the Burdett family and Messrs Barron & Son.

*Right*: Foremark Hall in 1840.

*Below*: Topiary in the Barron style.

The images of the gardens at Elvaston are a good reflection of Barron's style, which became common in a number of gardens and inspired many others, including those at Biddulph Grange. At Foremark the plans for the grounds included a lake surrounded by strategically placed foliage and a long, tree-lined, drive. The use of foliage to hide the size and views of the lake was typical of Barron's work, as was the creation of the 'Lime Drive', as it was marked in Barron's plans for Foremark. This avenue bears the hallmarks of Barron's earlier style as displayed at Elvaston, where he also elongated the route to the house via South Avenue and Vault Avenue, offering visitors a glimpse of the house as they approached through the gardens.

## Parkfield Cedars, Derbyshire

This large, detached brick villa was built in 1840 and remodelled and extended in the 1890s. The house was converted to a hospital in 1927. The grounds that had been laid out by Messrs Barron & Son have since been built over.

## Risley Hall, Derbyshire

Little remains of the original Risley Hall, which was the home of the Willoughby family, although it was described in *c.* 1710 as 'a large convenient building with good gardens, especially for fruit, and suitable to the estate, though no exact building'. In the 1870s some of the original garden buildings remained and were joined by a new (1790) house, which was further modified in 1890. In 1897, the grounds were relandscaped by William Barron & Son. Today the hall is a hotel and conference venue.

## Sennowe Park, Norfolk

Sennowe Park is a 66-hectare Norfolk estate comprising woodland, parkland and early twentieth-century formal gardens totalling some 1.5 hectares. The house, built by Thomas Wodehouse, dates back to 1774, as does the walled kitchen garden. During the 1850s William Barron was commissioned to layout the gardens and carry out work in the park. He employed woodsman and labourers and created a new entrance drive, similar to Elvaston, with rows of cedars and Douglas firs. Much of his work has been superseded by the remodelling of the house and gardens by George Skipper between 1905 and 1907. Skipper laid out extensive Italianate gardens and created a lake, various drives and lodges.

The photo included here was taken just before Skipper began work on the estate. It is, therefore, reasonable to assume that the urns, conservatory and topiary features were Barron's work – and they certainly seem typical of his style. It is also worth noting that the Register of Parks and Gardens cites Sennowe as having significant evidence of Barron's influence despite the later work by Skipper.

Sennowe Park. (Virginia & Charlie Temple-Richards)

# 13
# Public Works

The opening of Abbey Park. (Leicester City Council)

# Abbey Park, Leicester

In the 1870s the Abbey Meadows area on the edge of the city of Leicester was coming under increasing pressure from growing population density in the neighbouring Belgrave Gate, which was driven by the industrialisation of the canal network. This had led to concerns over public health as the frequent flooding in the area caused sewage contamination of the local water supply.

In 1879, the Leicester Corporation purchased Abbey Meadows from the Earl of Dysart as part of the Leicester Improvement Act, providing space to widen the river and reduce the risk of flooding. This project, which cost some £300,000 at the time, came with a commitment to subsequently turn the adjacent land included in the sale into 'a public park or recreation ground for the enjoyment of the inhabitants of Leicester'. In 1877, a competition was launched for the design of the park. Twenty-two designs were submitted and five were shortlisted and put on public display. There were prizes of 50, 30 and 20 guineas for first, second and third places respectively. No evidence remains of the amendments made to the design of the park before work began, though they seem to be few, with buildings put out as a separate contract and the inclusion of a southern entrance.

Messrs Barron & Son won the competition with a design that was largely the work of William Barron, supported by his son John. The project was initially viewed with public scepticism, with many claiming the site would be permanently waterlogged over the winter months. These concerns were to prove unfounded.

Over the next five years Barron designed the gardens, bandstands, rustic bridges and summerhouses, which still exist around the park. The pavilion and park lodges were designed by local architect James Tait. It has been suggested that Barron's design was inspired by the Arboretum in Derby, which had been laid out twenty-five years previously with a bandstand, boating lake, curved walks and a grand entrance.

As work progressed, the need to install a sewer beneath the park led to agreements with Barron to restrict the depth of the lake after trial holes for sewage drains revealed unstable strata below ground. The height of the water was increased, being pumped from further north on the river Soar. Rustic bridges designed by Barron and surrounded by low-growing evergreen foliage completed the scene around the lake.

Using an axial design, largely hidden by the arrangement of trees and circuitous paths, the park and lake were designed to create the illusion they were much larger. There was no single point in the park where visitors could acquire a full view of the lake. This approach follows a tradition that can be traced back to picturesque designers such as William Sawrey Gilpin and John Claudius Loudon.

The development of a new recreation ground for sports on the nearby Abbey Pastures allowed Barron to design a pleasure ground without having to consider football games. Abbey Park included an archery ground, lawn tennis, a bowling green and claimed to have space for cricket, though the final design would suggest otherwise, with sports facilities relegated to the peripheries. This left space in the core of the park for flower beds, extensive shrub varieties and viewing mounds.

Planting only began in 1880, with most of the plants supplied by Barron's own nursery in Borrowash, and a few more mature specimens being transplanted from

Barron's winning design for Abbey Park, Leicester. (Leicester City Council)

Barron's Tree Transplanter was capable of transporting trees a distance of 32 miles in a single journey.

Cropston reservoir. It is thought that this explains the size of some of the larger trees – particularly the London planes around the pavilion and bowling green. Barron undertook to replace any plants that died within twelve months of the planting scheme being delivered.

Barron also worked with the views the river afforded across to the ivy-covered abbey ruins, complemented by the extended river with a weir creating pools with two islands. This side of the park was much more open, with thicker planting to the west to eventually hide the factories. It is claimed extensively and in the Parks Restoration Plan that 'it is hard to overemphasise the significance of Abbey Park in the history of Landscape design in general, and in the development of urban parks in particular. It is not surprising that the park features prominently in most standard texts on Victorian gardens'. (Williamson, 1997)

The project was completed successfully in 1882 and the Prince of Wales and future Edward VII, together with his wife the Princess of Wales, were invited to officially open

Abbey Park. (Leicester City Council)

the park on Whit Monday in 1882. Thousands turned up to celebrate the event, lining the route from the railway station to Abbey Park.

Some areas have changed over the past century. The American Garden no longer exists and Tait's Pavilion was destroyed by fire in 1959 and replaced with a new pavilion contemporary to the period. In the 1980s a Japanese garden was created.

After the Dissolution of the Monasteries in 1538, a mansion was built on the site for the Marquis of Northampton. Acquired in 1613 by William Cavendish, 1st Earl of Devonshire, the house was used by Charles I after the siege of Leicester in 1645. It was subsequently torched by his soldiers, gutting the building. A charred stone window frame is still visible today.

The park continues to play an important part in the life of the city, and has been home to the City of Leicester Show and, during the 1980s, the Abbey Park Festival. Firework shows, a model railway, a pet corner and a boating lake continue to draw visitors to the park. Abbey Park is arguably one of the most complete examples of William Barron's public work.

It is worth noting for today's visitors that the park was extended by a further 32 acres on the other side of the River Soar in 1925 when the Earl of Dysart offered the Leicester Abbey grounds as a gift to the town council for use as a recreation ground and sports facilities This new part of the park incorporated the ruins of Cavendish House and the excavated remains of the abbey, which was founded by Robert le Bossu in 1138.

Opening of Abbey Park. (Leicester City Council)

Gardeners working near the Cavendish House ruins in Abbey Park, circa 1900. (Leicester City Council)

*Above, below and opposite*: Abbey Park. (Peter Robinson)

# Belper Cemetery, Derbyshire

In Victorian society evergreen trees were important symbols of piety and eternity, and yews had long been associated with sacred spaces. Such important symbolism aligned well with Barron's expertise and with his own moral and religious beliefs.

At Belper, Barron worked with Birmingham-based architect Edward Holmes. Holmes designed a lodge and twin mortuary chapels connected by a square tower and spire while Barron advised on the laying out of the cemetery. This work extended to land drainage, roads, planting and plots, drawing on ideas proposed by John Claudius Loudon that cemeteries had an important role to play in providing access to green space. Barron designed the landscape of the Belper Cemetery site as a picturesque arboretum rising from the banks of the Derwent, incorporating rockwork, mounds, numerous elegant evergreens and some deciduous trees. The land was consecrated on 16 June 1859 and has subsequently been extended several times.

*Above and opposite*: Belper Cemetery. (T. Liddle)

# Brunswick Park, Wednesbury

Work on Brunswick Park started in 1886 when former pit land was bought by Wednesbury Urban District Council for £3,000. Barron was commissioned to plan the park's layout and to supply the plants. After only a short period of work, the park opened in June 1887. It features an open lawn section with planting to the north and the original pit mound was landscaped and retained to the south. The pit mound is planted and has several paths converging at its peak.

# Locke Park, Barnsley

Locke Park is named after renowned railway engineer Joseph Locke (1805–60), who was brought up in Barnsley. Locke was an apprentice to George Stephenson and was the driver of *The Rocket* steam locomotive in the Rainhill Trials. Locke's widow, Phoebe, gave the 17-acre High Stile Field to the town on 24 April 1861 for a park in memory of her husband and as part of her husband's final wishes. Locke's former partner, John Edward Errington, arranged the initial layout of the park. Phoebe died in 1866 and in 1874 her sister Sarah McCreery donated a further 21 acres in her memory. An additional 1.5 acres were then donated by F. W. T. Vernon Wentworth of Wentworth Castle.

Sarah McCreery commissioned Richard Phené Spiers to design the park to include a tower that would act as both a memorial and viewing platform. A sketch plan by Spiers, dated 8 February 1875, shows a layout of winding paths linking the main areas of the park and culminating in a terrace just below the proposed tower. Landscaping was undertaken by William Barron & Son, recruited for their adventurous planting schemes with a mix of deciduous and evergreen trees and shrubs.

Locke Park Tower was formally opened on 20 October 1877, a little later than the opening of the additional park areas. Since this time a number of additional monuments have been introduced to the park, with a further 7-acre extension in 1914.

# Nottingham Road Cemetery, Derby

> *The grounds are tastefully laid out, and planted with evergreens and shrubs, under the able inspection of Mr Barron, of Elvaston Garden.*
>
> Glover (1858)

It is recorded that planting of the Nottingham Road Cemetery was undertaken by Mr Lee of Hammersmith, with advice from William Barron. Although the cemetery covers 32.2 hectares today, it was just 13 hectares when the Derby Burial Board (formed in 1853) consecrated the site in 1855 and became the first municipal burial site in the city.

*Above, right and overleaf*: Nottingham Road Cemetery. (T. Liddle)

# People's Park, Grimsby

Between 1860 and 1869 considerable effort was invested in campaigning for a park in Grimsby. The Great Grimsby Improvement Act of 1869 empowered the council to set aside part of the Grimsby West Marsh Area for a public park. This land was subsequently developed for other council projects, and it wasn't until 1881 when local landowner and MP Edward Heanage gave land to be developed as a public park. The land was drained and new drainage and roads were installed. The council also ordered the planting of around 700 trees, including approximately 120 elms, 240 limes, 120 sycamores, and 120 chestnuts, before the contract to lay out the grounds was awarded.

The competition to design the grounds received twenty-four entries, with Messrs Barron & Son winning the tender. Barron's *Semper Paratus* (Latin, meaning 'always ready') design drew together curves, open spaces and the ubiquitous evergreens with a double avenue of trees that had already been planted by the council. The plan was organised around a figure-of-eight circulation pattern. In the north the top loop contained the lake, mounds, and groups of tree and shrub planting, which created a series of unfolding views as the visitor walked along the path. The south loop remained flatter and more open to provide the cricket pitch and other sporting and recreational facilities.

The park formally opened in August 1883 and became a park of Special Historic Interest in 2000. It has already received funding from a number of schemes, including the Heritage Lottery Fund, to restore its original features.

# Queen's Park, Chesterfield

The park was designed to celebrate the Golden Jubilee of Queen Victoria in 1887 and its development lasted some six years. William Barron & Sons were asked to design the site, which covers 8 hectares just south of the town centre. When it was finally opened to the

Queen's Park, Chesterfield.

public in 1893, visitors found a lake, a perimeter walk and open space and lawns for sports. The park was further developed in the early 1920s and the 1950s and has several walks and a cricket pitch.

## Roath Park, Cardiff

Roath is the largest Cardiff suburb. At the heart of the district sits Roath Park, developed on land given by the Marquess of Bute in 1887, with two other benefactors – Lord Tredegar and the Lewis family – taking the total area of the park to 120 acres.

At the time it was the largest park in Wales. In 1889, William Barron & Son, were commissioned to design the park, which opened five years later on 20 June 1894, an event marked by the attendance of the marquess's heir on his thirteenth birthday.

Although it is unclear how much of Barron's design was implemented, the 30-acre artificial lake, recreation space, flower beds and botanical gardens certainly reflect Barron's overall approach to the creation of vistas and the use of water.

## Victoria Park, Sandwell

Victoria Park was laid out in 1898 on a former old mining site. Messrs Barron & Son's design was chosen from several submitted by different firms, with the landscaping work then carried out by T. Allsop of Tipton. The entrance lodge was built by J. Gittings of Tipton. The fencing was provided by Hill and Smith of Brierley Hill. Around 15,000 trees and shrubs were planted. After three years of work, the park was opened in 1901. The park now covers around 13 hectares, having been extended in the late 1900s.

## West Park, Macclesfield

A public park in Macclesfield was first proposed in 1850 and, after finally identifying suitable land, work to complete the park was undertaken very quickly, opening to the public on 2 October 1854, less than nine months after works began. Peel Park (as it was originally known) was the first sole public park commission undertaken by William Barron, which also makes it his first project on completion of works at Elvaston and the publication of *The British Winter Garden*. Thus, the park is historically much more important than is generally acknowledged.

Peel stood out from its contemporaries in having local political and aristocratic backing. The initial funding was raised through a campaign driven by trade unions, benefit societies and Chartists and had raised several hundred pounds, before being stalled by recession. When continued financial support dwindled, local gentry stepped in to renew the fundraising, together with the 5th Earl of Harrington.

It was through the 5th Earl of Harrington that Barron was asked to create the park, using techniques he had developed at Elvaston (which allegedly provided the earl with an opportunity to demonstrate that his indulgence at Elvaston was not driven by selfishness).

The original 16-acre site (today's park is much extended) was close to the top of a hill and offered commanding views over the local area. Barron's style is very evident at Peel: he approached the design of the pinetum as an aesthetic exercise, using the palette of evergreen hues he loved and employed at Elvaston.

At the request of John May, a local lawyer and financial supporter, the centre of the park was kept free for sport (which included both space for gymnastics and one of the largest bowling greens in the country). Around these spaces Barron created features that were clearly influenced by his work at Elvaston. The bowling green was surrounded by terraces with fences of *abor vitae* planted with Irish and golden yews. A series of artificial mounds exploited the views and echoed some of the features still to be found at Elvaston. One particularly large mound with steps to the top afforded some impressive views, while the park featured gothic entrance lodges, gates, and a pavilion.

A serpentine walk was designed across the park, with planting used to break up the views, featured ornamental trees while on the other side of the brook there was both a 'circuitous walk, cut through a plantation of ornamental trees' and a 'constitutional walk' that passed the gymnasium and completed a circuit of the park. The whole provided, according to the *Courier*, 'the shape of smiling lawns, graceful slopes and terraces, beautiful promenades and ornamental plantations'.

Much of the planting, especially around the edges and within the smaller plantations, was dominated by evergreens, especially yews to structure and divide the park into different areas.

As later at Abbey Park in Leicester, Barron believed it was essential that trees performed two functions: to provide shelter and 'for effect and depth to pictorial views which may be produced afterwards, to plant thickly all backgrounds as a primary operation'.

## Worcester Pleasure Grounds, Worcestershire

Little remains of the Barron-designed Worcester Pleasure Grounds, which later became part of the Worcester Arboretum. In turn, the Arboretum now gives its name to a part of the city where only a few traces remain of the former tennis courts, ice-skating rink, gymnasium, Turkish baths and other features. The pleasure grounds were not dissimilar to the approach Barron took at Abbey Park and West Park, creating an elaborate plan with a basic symmetry, hidden by curved walks, geometric flower beds and picturesque planting.

The pleasure grounds also featured broderie parterres surrounded by serpentine walks, terraces, promenades, flower beds, a bowling green, cricket pitch and archery butts, all enclosed in elaborate iron gates and ornamental palisading. Other features included a glass pavilion, central fountain and an elm avenue.

Just five years after opening, the Worcester Public Pleasure Grounds Company had run out of money and, with debts of £6,000 (despite holding numerous public events including fireworks shows and gymnastics), it was declared bankrupt. Although the local council proposed taking on the park, the motion was defeated at a council meeting and the land was subsequently sold for housing.

Worcester Pleasure Grounds.

# 14

# Other Messrs Barron & Sons Works

For a firm that survived for almost a century there are many more properties and parks designed and developed by William Barron and Messrs Barron & Sons. Many of these will be lost to the records of time, but this brief list identified some further projects where less information is available:

- Alvaston Park, Derby – opened to the public in 1913, after William Curzon of Breedon Hall set aside 12 hectares (30 acres) of land.
- Borrowash Golf Course, Derbyshire – laying out of the links.
- Fressingfield, No. 116 Blagreaves Lane, Derby – the gardens were relandscaped by William Barron & Sons both in 1913 and 1924.
- Hall Leys Park, Matlock – planting around the bowling green.
- Littleover Old Hall, Derbyshire – gardens and pleasure grounds (now mainly built over) landscaped by William Barron & Son after 1898 and relandscaped in 1934 by Barron after the property was sold to Harold Walker in 1934 with gardens reduced to 10 acres.
- Mickleover Manor, Derbyshire – the grounds were laid out by William Barron c. 1855. The pleasure grounds were once 17 acres, walled round in brick, but now reduced to 7 acres. An original stable block survives.
- Park Field, Duffield Road, Derby (since demolished/developed) – between 1857 and 1861 William Barron was called in to landscape the pleasure grounds. There are fragments of garden ornaments lying on the ground, as well as the base of a small building and a tennis court.
- Tissington Hall, Derbyshire – carried out unspecified works in 1913.
- Victoria Park, Widnes – a late nineteenth-century public park opened in 1900 to commemorate Queen Victoria's Diamond Jubilee. Features include shrubberies, display glasshouses, carpet bedding, a war memorial, tennis courts, bowling greens and a new bandstand replacing the original.
- Whittaker Park, Rawtenstall – the park and its associated house were created in 1840 and gifted to Rawtenstall in 1902 when the house became a museum and the grounds a public park. The site was designed by John Barron, and is landscaped, with formal gardens.

# 15

# Elvaston Today

The gardens retain many of the features that Barron would still recognise. Some highlights are presented here, which both illustrate this point and provide a catalogue of design features that are synonymous with Barron's work.

There are two Wellingtonia at Elvaston, situated at the end of Church and School drives. Part of the redwood family, these graceful giants were two of the first specimens of their type brought to England in the 1800s. (T. Liddle)

Bird Cottage was an elaborate bird-themed topiary devised by William Barron. It sits at the end of Church Drive at Elvaston. Although its shape has changed slightly, it is still possible to make out the shapes of bird heads and tails. (T. Liddle)

Barron's rockwork appears and disappears as visitor walk through the grounds. It is with delight that visitors find these unusual shapes and eyecatchers, designed to provide very deliberately chosen vistas at every turn. (T. Liddle)

The Kitchen Garden still grows a variety of different produce thanks to the wonderful fruit trees trained around the walls. The glass houses in the garden are in a state of disrepair. It is a future aspiration of the Trust's to renovate them so they can, once again, be home to special varieties of fruit and vegetables. The Kitchen Garden. (T. Liddle)

Even on an overcast day Elvaston is a wonderful place to visit. Here you can see the earlier brick façade of the original H-shaped Elizabethan house, which was incorporated into the 4th Earl's nineteenth-century renovation. Inside the house, the remodelling created a secret room between two floors, indicated by what appears to be the top floor window in this Elizabethan corner. (T. Liddle)

*Above*: Around the estate there is a well-loved bridlepath, providing a beautiful route for equestrian, cycling, and walking visitors alike. The path bounds the current showground, once the Harrington's polo field. Barron improved the drainage on this side of the estate to ensure the ground was suitable for sport. (T. Liddle)

*Opposite*: The grand staircase sits at the heart of the castle and was originally decorated with portraits from floor to ceiling. Visitors would be greeted in the Hall of Fair Star and then wined and dined on the ground floor. They would then be swept up to the first- and second-floor guest rooms for a night's rest. (T. Liddle)

*Above*: The majestic south drive was once the entry point to the estate for visitors. The 4th Earl wanted his guests to be amazed by the grandeur of the estate and, like many stately homes of the time, a wide tree-lined avenue was developed to impress guests. Although aligned with the south front of the house, Barron created a yew tree crown which prevented direct access. Guests, after glimpsing the crenelations, would turn right, passing other gardens, which became increasingly dark and gloomy before emerging into the light at the southern side of the house. Later garden visitors were greeted here with a cast-iron arbour with a 'welcome to Elvaston banner' atop. (T. Liddle)

*Opposite*: Once the beating heart of the castle, the Gothic kitchen has been reopened to the public by the Elvaston Castle and Gardens Trust. While it has deteriorated from its former glory, the kitchen is a still fascinating space with multiple oven ranges. Some of the earliest gas rings (potentially the earliest) are ornamented with a family crest and French motto translating to 'A cook is a divine mortal'. This was a show kitchen to impress visitors. (T. Liddle)

Alongside the Hall of Fair Star, the walled garden today plays host to weddings. A beautiful, calm space with seasonal planting, this enclave is also home to a sensory garden. This was once the heart of Barron's thriving kitchen garden, producing some of the best fruits in Europe. The Walled Garden. (T. Liddle)

# Further Reading and References

- Barron, William, *The British Winter Garden: Being a Practical Treatise on Evergreens; Showing Their General Utility in the Formation of Garden and Landscape Scenery* (London: Bradbury & Evans, 1852)
- Batey, Mavis, 'Edward Cooke: Landscape Gardener', *Garden History*, 6 (1978), pp. 18–24
- Elliott, B., *Victorian Gardens* (London: Batsford Ltd, 1986).
- Hayden, Peter, *Biddulph Grange: A Victorian Garden Re-discovered* (London: National Trust)
- Kemp, E., *Description of the Gardens at Biddulph Grange* (London: Bradbury & Evans, 1862; 1989)
- The Gardens Trust Stable, *Cultural and Historical Geographies of the Arboretum* (2007), pp. 129–48
- Barron, W. and Nineteenth-Century British Arboriculture, *Evergreens in Victorian Industrializing Society*, Paul Elliott, Charles Watkins and Stephen Daniels, from *Garden History* (2007), Vol. 35

# Acknowledgements

We would like to pass on our appreciation to the following contributors, organisations and proofreaders:

Matthew Allen, Derbyshire County Council
Grahame Appleby, Leicester City Council
Andrew Baker, Trustee, The Elvaston Castle and Gardens Trust
Stefan Cabaniuk, Leicester City Council
Lucy Bamford, Derby Museums
Rachel Fannen, Chesterfield Museum
Katherine Harrington, Royal Botanic Gardens, Kew
Margie Hoffnung, The Gardens Trust
Su Hurrell, Betley Court
Samantha Harvey, Derby College
Dr David Marsh, The Gardens Trust
Kevin Martin, Royal Botanic Gardens, Kew
Adrienne McStocker, Volunteer at Broomfield Hall
Rob Price, Advisory Panel, The Elvaston Castle and Gardens Trust
Leonie Paterson, Royal Botanic Gardens Edinburgh
Becky Sheldon, Derbyshire County Council
Connor Stait, Amberley Publishing
Virginia and Charlie Temple-Richards, Sennowe Park
Ian Waller, Chesterfield Borough Council
Vaughn Wheatley, Derbyshire County Council

# About the Trust and the Authors

## The Elvaston Castle and Gardens Trust

The Elvaston Castle and Gardens Trust (ECGT) is a charity working to ensure that Elvaston has a sustainable future. Working alongside Derbyshire County Council, the Trust is developing plans which will to reverse decades of decline, ensuring that future generations are able to enjoy this special place.

*You can find out more about Elvaston, the Trust and how to support our work by visiting www.futureelvaston.co.uk.*

# Tamsin Liddle

A lifelong member of the National Trust, Tamsin Liddle fell in love with historic houses and gardens at an early age. She is a senior leader in the aviation industry, with expertise in purchasing, programme management and business transformation. An amateur photographer, she enjoys travelling, learning, walking and creating. Tamsin joined the Elvaston Castle & Gardens Trust in 2019, was appointed its Vice Chair in 2020 and is dedicated to making sure that Elvaston has a bright and secure future.

# Dr Peter Robinson

Dr Peter Robinson has been a regular visitor to Elvaston since a school visit when he was four years old. He is Head of the Centre for Tourism and Hospitality Management at Leeds Beckett University, a Director of the Institute of Travel and Tourism, a Principal Fellow of the Higher Education Academy and the Yorkshire and Humber representative for the Tourism Management Institute. Peter is a renowned expert in tourism with an international publishing and media profile. He is also an amateur photographer and engineer, a keen walker and cyclist and Chair of the Elvaston Castle and Gardens Trust.